Lecture Notes in Computer Science 8380

Commenced Publication in 1973
Founding and Former Series Editors:
Gerhard Goos, Juris Hartmanis, and Jan van Leeuwen

Editorial Board

David Hutchison
 Lancaster University, UK
Takeo Kanade
 Carnegie Mellon University, Pittsburgh, PA, USA
Josef Kittler
 University of Surrey, Guildford, UK
Jon M. Kleinberg
 Cornell University, Ithaca, NY, USA
Alfred Kobsa
 University of California, Irvine, CA, USA
Friedemann Mattern
 ETH Zurich, Switzerland
John C. Mitchell
 Stanford University, CA, USA
Moni Naor
 Weizmann Institute of Science, Rehovot, Israel
Oscar Nierstrasz
 University of Bern, Switzerland
C. Pandu Rangan
 Indian Institute of Technology, Madras, India
Bernhard Steffen
 TU Dortmund University, Germany
Demetri Terzopoulos
 University of California, Los Angeles, CA, USA
Doug Tygar
 University of California, Berkeley, CA, USA
Gerhard Weikum
 Max Planck Institute for Informatics, Saarbruecken, Germany

Andreas Kerren   Helen C. Purchase
Matthew O. Ward (Eds.)

Multivariate Network Visualization

Dagstuhl Seminar #13201
Dagstuhl Castle, Germany, May 12-17, 2013
Revised Discussions

Volume Editors

Andreas Kerren
Linnaeus University, Faculty of Technology (FTK)
Department of Computer Science
Vejdes Plats 7, 351 95 Växjö, Sweden
E-mail: kerren@acm.org

Helen C. Purchase
University of Glasgow, School of Computing Science
18 Lilybank Gardens, Glasgow G12 1RZ, UK
E-mail: hcp@dcs.gla.ac.uk

Matthew O. Ward
Worcester Polytechnic Institute, Department of Computer Science
100 Institute Road, Worcester, MA 01609-2280, USA
E-mail: matt@cs.wpi.edu

ISSN 0302-9743 e-ISSN 1611-3349
ISBN 978-3-319-06792-6 e-ISBN 978-3-319-06793-3
DOI 10.1007/978-3-319-06793-3
Springer Cham Heidelberg New York Dordrecht London

Library of Congress Control Number: 2014937318

LNCS Sublibrary: SL 3 – Information Systems and Application, incl. Internet/Web and HCI

© Springer International Publishing Switzerland 2014

This work is subject to copyright. All rights are reserved by the Publisher, whether the whole or part of the material is concerned, specifically the rights of translation, reprinting, reuse of illustrations, recitation, broadcasting, reproduction on microfilms or in any other physical way, and transmission or information storage and retrieval, electronic adaptation, computer software, or by similar or dissimilar methodology now known or hereafter developed. Exempted from this legal reservation are brief excerpts in connection with reviews or scholarly analysis or material supplied specifically for the purpose of being entered and executed on a computer system, for exclusive use by the purchaser of the work. Duplication of this publication or parts thereof is permitted only under the provisions of the Copyright Law of the Publisher's location, in ist current version, and permission for use must always be obtained from Springer. Permissions for use may be obtained through RightsLink at the Copyright Clearance Center. Violations are liable to prosecution under the respective Copyright Law.
The use of general descriptive names, registered names, trademarks, service marks, etc. in this publication does not imply, even in the absence of a specific statement, that such names are exempt from the relevant protective laws and regulations and therefore free for general use.
While the advice and information in this book are believed to be true and accurate at the date of publication, neither the authors nor the editors nor the publisher can accept any legal responsibility for any errors or omissions that may be made. The publisher makes no warranty, express or implied, with respect to the material contained herein.

Typesetting: Camera-ready by author, data conversion by Scientific Publishing Services, Chennai, India

Printed on acid-free paper

Springer is part of Springer Science+Business Media (www.springer.com)

Preface

During May 12–17, 2013, a seminar on "Information Visualization – Towards Multivariate Network Visualization" (no. 13201) took place at the International Conference and Research Center for Computer Science, Dagstuhl Castle, Germany. The center was initiated by the German government to promote computer science research at the international level. It seeks to foster dialog among the research community, advance academic education and professional development, and transfer knowledge between academia and industry.

Information visualization (InfoVis) is a research area that focuses on the use of visualization techniques to help people understand and analyze data as well as relations between data. While related fields such as scientific visualization involve the presentation of data that have some physical or geometric correspondence (for example, climate patterns, molecular formations, transport networks), InfoVis centers on abstract information without such correspondences, i.e., it is not possible to map this information into the physical world in most cases. Examples of such abstract data are symbolic, tabular, networked, hierarchical, or textual information sources—for example, genealogies, demographic data of a population, or financial trends.

The goal of this third Dagstuhl Seminar on Information Visualization was to bring together theoreticians and practitioners from InfoVis, HCI, and graph drawing with a special *focus on multivariate network visualization*, i.e., on graphs where the nodes and/or edges have additional (multidimensional) attributes. The integration of multivariate data into complex networks and their visual analysis is one of the big challenges not only in visualization, but also in many application areas. Thus, in order to support discussions related to the visualization of real-world data, we also invited researchers from selected application areas, especially bioinformatics, social sciences, and software engineering. The unique "Dagstuhl climate" ensured an open and undisturbed atmosphere to discuss the state-of-the-art, new directions, and open challenges of multivariate network visualization.

This book is the outcome of Dagstuhl Seminar no. 13201. It documents and extends the findings and discussions of the various sessions in detail. During the last day of the seminar, the most important topics for publication were identified and assigned to interested participants. The resulting author groups worked together to write book chapters on the chosen topics.

We would like to thank all participants of the seminar for the lively discussions and contributions during the seminar as well as the scientific directorate of Dagstuhl Castle for giving us the possibility to organize this event. The abstracts and presentation slides can be found on the Dagstuhl website for this seminar[1]. There is an online document that reports on all activities during the seminar[2]. We are also grateful to all the authors for their valuable time and contributions to the book. Last but not least, the seminar and thereby this book would not have been possible without the great help of the staff of Dagstuhl Castle. We would like to acknowledge all of them for their assistance.

January 2014

Andreas Kerren
Helen C. Purchase
Matthew O. Ward

---

[1] http://www.dagstuhl.de/13201
[2] http://dx.doi.org/10.4230/DagRep.3.5.19

List of Contributors

James Abello
   Rutgers University, DIMACS Center for Discrete Mathematics and Theoretical Computer Science, USA
   abello@dimacs.rutgers.edu

Daniel Archambault
   Swansea University, Department of Computer Science, UK
   d.w.archambault@swansea.ac.uk

Katy Börner
   Indiana University Bloomington, Department of Library and Information Science, USA
   katy@indiana.edu

Stephan Diehl
   University of Trier, Department of Computer Science, Germany
   diehl@uni-trier.de

Tim Dwyer
   Monash University Clayton, Caulfield School of Information Technology, Australia
   tim.dwyer@monash.edu

Niklas Elmqvist
   Purdue University, School of Electrical & Computer Engineering, USA
   elm@purdue.edu

Jean-Daniel Fekete
   INRIA, France
   Jean-Daniel.Fekete@inria.fr

Liang Gou
　IBM Research Almaden, USA
　lgou@us.ibm.com

Hans Hagen
　University of Kaiserslautern, Department of Computer Science, Germany
　hagen@cs.uni-kl.de

Danny Holten
　SynerScope, Eindhoven, The Netherlands
　danny.holten@synerscope.com

Christophe Hurter
　École Nationale de l'Aviation Civile, Interactive Computing Laboratory, France
　christophe.hurter@enac.fr

T.J. Jankun-Kelly
　Mississippi State University, Department of Computer Science and Engineering, USA
　tjk@acm.org

Jessie Kennedy
　Edinburgh Napier University, School of Computing, UK
　j.kennedy@napier.ac.uk

Andreas Kerren
　Linnaeus University Växjö, Department of Computer Science, Sweden
　kerren@acm.org

Stephen Kobourov
　University of Arizona, Department of Computer Science, USA
　kobourov@cs.arizona.edu

Oliver Kohlbacher
　University of Tübingen, Center for Bioinformatics,
　Quantitative Biology Center, Department of Computer Science,
　and Faculty of Medicine, Germany
　oliver.kohlbacher@uni-tuebingen.de

Kwan-Liu Ma
　University of California at Davis, Department of Computer Science, USA
　ma@cs.ucdavis.edu

Silvia Miksch
Vienna University of Technology, Institute of Software Technology and Interactive Systems, Austria
miksch@ifs.tuwien.ac.at

Chris Muelder
University of California at Davis, Department of Computer Science, USA
cwmuelder@ucdavis.edu

Martin Nöllenburg
Karlsruhe Institute of Technology, Institute of Theoretical Informatics, Germany
noellenburg@kit.edu

A. Johannes Pretorius
University of Leeds, School of Computing, UK
a.j.pretorius@leeds.ac.uk

Helen C. Purchase
University of Glasgow, School of Computing Science, UK
hcp@dcs.gla.ac.uk

Jonathan C. Roberts
Bangor University, School of Computer Science, UK
j.c.roberts@bangor.ac.uk

Falk Schreiber
Martin Luther University Halle-Wittenberg, Institute of Computer Science, Germany; Leibniz Institute of Plant Genetics and Crop Plant Research (IPK), Germany; and Monash University Melbourne, Clayton School of Information Technology, Australia
schreibe@ipk-gatersleben.de

John T. Stasko
Georgia Institute of Technology, School of Interactive Computing, USA
stasko@cc.gatech.edu

Alexandru C. Telea
University of Groningen, Institute Johann Bernoulli, The Netherlands
a.c.telea@rug.nl

Jarke J. van Wijk
Eindhoven University of Technology, Department of Mathematics and Computer Science, The Netherlands
vanwijk@win.tue.nl

Tatiana von Landesberger
Technische Universität Darmstadt, Department of Computer Science, Germany
tatiana.von-landesberger@gris.tu-darmstadt.de

Matthew O. Ward
Worcester Polytechnic Institute, Department of Computer Science, USA
matt@cs.wpi.edu

Chris Weaver
University of Oklahoma, School of Computer Science, USA
weaver@cs.ou.edu

Michael Wybrow
Caulfield School of Information Technology, Monash University Caulfield, Australia
Michael.Wybrow@monash.edu

Kai Xu
Middlesex University London, School of Science and Technology, UK
k.xu@mdx.ac.uk

Jing Yang
University of North Carolina at Charlotte, Computer Science Department, USA
Jing.Yang@uncc.edu

Dirk Zeckzer
Leipzig University, Institute of Computer Science, Germany
zeckzer@informatik.uni-leipzig.de

Michelle X. Zhou
IBM Research Almaden, USA
mzhou@us.ibm.com

Björn Zimmer
Linnaeus University Växjö, Department of Computer Science, Sweden
bjorn.zimmer@lnu.se

Contents

Preface V

List of Contributors VII

1 **Introduction to Multivariate Network Visualization** 1
  Andreas Kerren, Helen C. Purchase, Matthew O. Ward
  1.1 Multivariate Networks: Definitions and
      Terminology 2
  1.2 Existing Visualizations 3
  1.3 Outline of This Book 6
  References 7

Part I: Application Domains – Characteristics and Challenges

2 **Multivariate Networks in Software Engineering** 13
  Stephan Diehl, Alexandru C. Telea
  2.1 Aims and Scope 13
      2.1.1 History and Definitions 14
      2.1.2 Importance 14
  2.2 Data Characteristics 15
      2.2.1 Entities 15
      2.2.2 Relations 15
      2.2.3 Attributes 16
      2.2.4 Software as Multivariate Time-Dependent Graphs .. 17
      2.2.5 Reference Implementation 17
      2.2.6 Software Data *vs.* other InfoVis Domains 19
  2.3 Applications 21
      2.3.1 Structure Visualization 21
      2.3.2 Behavior Visualization 26
      2.3.3 Evolution Visualization 28
  2.4 Challenges and Future Directions 32
  References 34

3 Multivariate Social Network Visual Analytics 37
Chris Muelder, Liang Gou, Kwan-Liu Ma, Michelle X. Zhou
- 3.1 Data Characteristics 37
  - 3.1.1 Traditional Data Collection 38
  - 3.1.2 Data Collection from Online Social Media Networks ... 38
- 3.2 Task Characteristics 39
  - 3.2.1 Understanding Social Network Nodes 39
  - 3.2.2 Understanding Social Network Links 39
  - 3.2.3 Understanding Social Networks 39
- 3.3 Examples of Technologies 40
  - 3.3.1 Clustering 40
  - 3.3.2 Network Centralities 41
  - 3.3.3 Centrality Derivatives 41
  - 3.3.4 Traditional Network Layouts 42
  - 3.3.5 Improved Network Layouts 43
  - 3.3.6 Multivariate Social Networks 44
  - 3.3.7 Dynamic Social Networks 46
  - 3.3.8 Egocentric Approaches 47
  - 3.3.9 TreeNetViz: Revealing Patterns of Networks with Hierarchical Attributes 47
  - 3.3.10 SocialNetSense: Making Sense of Multivariate Social Networks 50
  - 3.3.11 Summary 54
- 3.4 Challenges and Future Directions 55
- References 56

4 Multivariate Networks in the Life Sciences 61
Oliver Kohlbacher, Falk Schreiber, Matthew O. Ward
- 4.1 Characteristics of Data and Tasks 62
  - 4.1.1 Types of Biological Networks 62
  - 4.1.2 Data Mapping and Multivariate Networks 63
- 4.2 Use Cases 66
  - 4.2.1 Signaling 66
  - 4.2.2 Genetic Linkage 66
  - 4.2.3 Relationship Discovery Based on Document Analysis 67
  - 4.2.4 Gene Regulation and Transcriptome Data 68
- 4.3 Challenges 69
  - 4.3.1 Scale 70
  - 4.3.2 Uncertainty/Ambiguity 70
  - 4.3.3 Heterogeneity 71
  - 4.3.4 Interactivity 71
  - 4.3.5 Standardization 71
- 4.4 Summary and Conclusions 72
- References 72

Part II: Topics in Multivariate Network Research

5 Tasks for Multivariate Network Analysis 77
A. Johannes Pretorius, Helen C. Purchase, John T. Stasko
  5.1 Entities and Properties 78
  5.2 Tasks 79
        5.2.1 General Task Taxonomy 80
        5.2.2 Tasks for Multivariate Network Analysis 83
  5.3 Discussion 92
  5.4 Conclusion 94
  References 95

6 Interaction in the Visualization of Multivariate Networks 97
Michael Wybrow, Niklas Elmqvist, Jean-Daniel Fekete, Tatiana von Landesberger, Jarke J. van Wijk, Björn Zimmer
  6.1 Background 98
  6.2 Classification of Interactions 101
        6.2.1 View-Level Interactions 102
        6.2.2 Visual Structure-Level Interactions 104
        6.2.3 Data-Level Interactions 106
  6.3 Exemplars 109
        6.3.1 GraphDice 110
        6.3.2 GraphTrail 111
        6.3.3 State Transition Networks 112
        6.3.4 Parallel Node-Link Bands 113
  6.4 Recommendations and Guidelines 115
        6.4.1 Learnability 115
        6.4.2 Flexibility 116
        6.4.3 Robustness 117
  6.5 Challenges and Vision 118
  References 121

7 Novel Visual Metaphors for Multivariate Networks 127
Jonathan C. Roberts, Jing Yang, Oliver Kohlbacher, Matthew O. Ward, Michelle X. Zhou
  7.1 Background 128
        7.1.1 Semantically Rich Data 128
        7.1.2 Where Ideas Come From? 129
        7.1.3 The Ideation Process 131
        7.1.4 The Visual Mapping Process 132
  7.2 Classes of Metaphors 135
        7.2.1 Nature-Inspired 135
        7.2.2 Non-physical 140

	7.3	Man-Made..	141
	7.4	Visualization-Inspired.....................................	142
	7.5	Proposed New Ideas.......................................	144
		7.5.1 Graphs/Networks	145
		7.5.2 Hierarchies	145
	7.6	Summary and Conclusions.................................	147
		References	147

8 Temporal Multivariate Networks 151
Daniel Archambault, James Abello, Jessie Kennedy,
Stephen Kobourov, Kwan-Liu Ma, Silvia Miksch,
Chris Muelder, Alexandru C. Telea

	8.1	Definitions...	151
		8.1.1 Structure, Behavior, and Evolution	152
		8.1.2 Formal Definitions of Temporal Multivariate Networks...	153
	8.2	Refining Our Models and Definitions for Time	154
	8.3	Survey of Representations and Algorithms	156
		8.3.1 Static Graph Layouts	157
		8.3.2 Dynamic Graph Layouts	158
		8.3.3 Animation versus Small Multiples	161
		8.3.4 Mental Map Preservation	162
		8.3.5 Alternative Representations.......................	162
		8.3.6 Static Temporal Plots	163
		8.3.7 Dynamic Graph Analytics	164
	8.4	Applications to Software Engineering	165
	8.5	Open Problems..	167
		8.5.1 Attribute Dimensionality	167
		8.5.2 Capturing Patterns	168
		8.5.3 Data Size	168
	8.6	Summary and Conclusions................................	169
		References	169

9 Heterogeneous Networks on Multiple Levels............. 175
Falk Schreiber, Andreas Kerren, Katy Börner,
Hans Hagen, Dirk Zeckzer

	9.1	Formal Description of Used Data Structures	177
	9.2	Application Domains......................................	179
		9.2.1 Life Sciences / Biology	179
		9.2.2 Social Science.....................................	185
		9.2.3 Software Engineering	190
	9.3	Visualization..	196
		9.3.1 Approaches for Networks at Multiple Levels	197
		9.3.2 Challenges and Future Directions..................	200
		References	201

10 Scalability Considerations for Multivariate Graph Visualization 207
T.J. Jankun-Kelly, Tim Dwyer, Danny Holten,
Christophe Hurter, Martin Nöllenburg, Chris Weaver, Kai Xu

- 10.1 Limits of Visualization 208
  - 10.1.1 Limits of Visual Acuity 208
  - 10.1.2 Cognitive Limits 210
  - 10.1.3 Leveraging the Graphics Card (GPU) 211
- 10.2 Design Strategies for Scalable Multivariate Graph Visualization 215
  - 10.2.1 Data Transformation and Reduction 217
  - 10.2.2 Visual Mapping 222
  - 10.2.3 View Transformation 223
- 10.3 Studies on Scalability in Graph Visualization 224
  - 10.3.1 Data Transformation and Reduction 224
  - 10.3.2 Visual Mapping 225
  - 10.3.3 Navigation and Interaction 227
- 10.4 Challenges and Future Directions 229
- References 229

Author Index 237

1

Introduction to Multivariate Network Visualization

Andreas Kerren, Helen C. Purchase, and Matthew O. Ward

Information Visualization (InfoVis) research focuses on the use of techniques to help people understand and analyze data. In particular, it considers how abstract data (i.e., without correspondence to the physical world) can best be visually represented. A variety of different abstract data types are addressed in InfoVis research (e.g., numerical, ordinal, categorical [29]), all of which can be arranged in different ways: for example, in linear, tabular, or network form. Common representations of statistical data (e.g., pie charts, bar charts, or scatter plots) are all visualizations of abstract numerical data.

A "multivariate network" (MVN) is an abstract data type that provides particular challenges for the information visualization research community. It permits the representation of complex relational data (stored in the form of a network) as well as the association of attributes with that data. The attributes themselves may use a range of different abstract data types.

A MVN therefore consists of a set of objects, each of which has information associated with it. In addition, the objects are connected to each other in a network that represents the relationship between the objects. Further complexity is added when information is also associated with the inter-object relationships themselves. For example, a social network representation may consist of people (the objects), each of which has information associated with them (their age and post code). Friendship relationships between the people form the network, with additional information about each friendship between two people being stored (e.g., the last time they communicated with each other or how long they have known each other).

In InfoVis terms, the objects in the network are called *nodes* or *vertices*, the relationships between the objects are called *links* or *edges*, the network is mostly called a *graph*, and the information associated with the objects and relationships are called *attributes*, *features*, *dimensions*, or *properties*.

MVNs prove particularly challenging for InfoVis researchers because of the wealth, richness and variety of the information that can be stored in them. Any number of attributes can be associated with nodes and edges, and nodes can be associated with any number of other nodes. Depicting all this information in visual form so as to help people understand it is a clear challenge. In many cases, given the limits of human perceptual and cognitive abilities, it is impossible to clearly show all the information at once in a useful form. Such problems may be addressed by prior knowledge of the user tasks to be performed, by providing facilities for interacting with the data, or by a clever choice of representation. These problems are of course made even more complex if the information changes over time, if there is more than one network, or if the network is particularly large.

Despite the challenges of depicting MVNs, their existence is common, and they are (and have been) used to represent abstract domain knowledge for many years (for example in software engineering, biology, evolution, environmental sciences, meteorology, and sociology). The popular educational tool of "concept mapping" demonstrates the wide range of applications to which MVNs can be applied [23]. Studying ways in which computational techniques may be used to help with the effective visualization of MVNs is therefore an area of study with wide applicability as well as the potential to be highly useful in other research areas.

In this chapter, we define MVNs formally, introduce and classify existing related InfoVis research, and provide a brief summary of the rest of the book.

1.1 Multivariate Networks: Definitions and Terminology

A *(simple) graph* $G = (V, E)$ consists of a finite set of vertices (or nodes) V and a set of edges $E \subseteq \{(u, v) | u, v \in V, u \neq v\}$. Based on this, a variety of general graph properties and characteristics can be found in the literature; the most important ones are introduced in the following list (cf. also the book chapter [18] or the books [7, 14]).

- An edge $e = (u, v)$ with $u = v$ is called a *self-loop*.
- If an edge e exists several times in E then it is called a *multiple edge*.
- A *simple graph* has no self-loops and no multiple edges.
- The *neighbors* of a node v are its adjacent nodes.
- The *degree* of a node v is the number of its neighbors.
- A *directed graph* (or digraph) is a graph with directed edges, i.e., (u, v) are ordered pairs of nodes.
- A directed graph is called *acyclic* if it has no directed cycles, i.e., there is no directed path where the same node is visited twice.
- A graph is *connected* if there is a path between u and v for each pair (u, v) of nodes.
- A graph is *planar* if it can be drawn in the 2D plane without intersections of edges (*edge crossings*).

In contrast to the above definition of a simple graph, a *multivariate network* N consists of an underlying graph G plus n additional attributes

$A = \{A_1, \ldots, A_n\}$ that are attached to the nodes (and/or edges). For node attributes, A_i represents a column in a table of attributes $A = (a_{ji})\,(j = 1 \ldots |V|\,; i = 1 \ldots n)$ and contains one attribute value per node (similar definition for edges). Thus, $a^u = (a_{u1}, \ldots, a_{un})$ describes all attribute values for node u given that there is no missing data.

Also in network theory, researchers have developed a set of useful measurements and metrics that can be used to get an impression about the most important characteristics of the graph topology such as central actors in a social network [22]. Those measures can also be applied to multivariate networks. *Community analysis* based on specific clustering techniques is one such approach. Another example are so-called *network centralities*, i.e., measures that quantify how important a node or edge in the network is. More formal, a network centrality C is a function that assigns a value $C(u)$ to a node $u \in V$ of a given graph $G = (V, E)$. This function supports centrality comparisons according to their importance, i.e., u is more important than v iff $C(u) > C(v)$ [8, 20]. A simple example of a network centrality is the degree of a node in an undirected graph.

1.2 Existing Visualizations

In the following, we briefly highlight the most important visualization techniques for arbitrary multivariate graphs/networks, i.e., we do not consider special cases such as the visualization of hierarchies (trees) or directed acyclic graphs (DAGs). The interested reader is referred to the vast amount of literature on these topics as for instance [11, 30].

Before we continue our discussion on multivariate network visualizations, we turn the reader's attention to standard techniques for the visualization of multivariate data itself. Multivariate (or multidimensional) data sets can mostly be described as data tables with n data objects and m attributes/features, i.e., for each object exists an attribute vector with m dimensions. The attribute values can be classified—for instance—into nominal, ordinal, or quantitative. In practice, we often have a large amount of data objects and many attributes with different types. Finding a suitable visual representation is thus challenging, and the right choice might depend on further parameters like application domain, integration into a larger visualization environment, or support of specific interaction techniques. In general, visual mappings for multivariate data can roughly be categorized into *point-based approaches* (e.g., scatterplot matrices [6], projection methods like MDS [21, 34], etc.), *axis-based approaches* (parallel coordinate plots [10], Kiviat diagrams [3], etc.), *icon-based approaches* (Chernoff faces [5], stick figures [24], etc.), and *pixel-based approaches* (e.g., recursive patters [15], pixel bar charts [16], etc.). There are many good textbooks that provide a good overview of those methods; we recommend the books of Spence [29], Kerren et al. [19], and Ward et al. [31].

A View on Graph Drawing

Traditional graph drawing (GD) methods compute a 2D/3D layout of the nodes and the edges, mainly based on *node-link diagrams* [32]. They play a fundamental role in network visualization. Particular graph layout algorithms can give an insight into the topological structure of a network if properly chosen and implemented. The graph readability is affected by quantitative measurements called *aesthetic criteria* [7]. Thus, graph drawing generally deals with the ways of drawing graphs according to the set of predefined aesthetic criteria [4]. These criteria are often contradictory, and problems which aim to optimize the criteria are often NP-hard. Therefore, many GD algorithms are heuristics. For the sake of completeness, we want to note that there are also so-called space-filling methods that try to solve some conceptual problems of node-link diagrams, such as the high space consumption or edge crossings. *Matrix displays* fall into this category (like the approach proposed in [1]). These visualizations represent a graph directly via its adjacency matrix, where a matrix element (i, j) represents the existence of an edge between the two nodes i and j. A disadvantage of matrices is that the perception of the graph topology is depending of the node order in the matrix.

In both visual representations (i.e., node-link and matrix displays), multivariate data can be integrated in various ways. For instance in node-link diagrams, multivariate glyphs that replace the node representations (usually a dot or circle) can be used to show data attached to nodes; edge attributes can be represented by different link colors, thicknesses, labels, or edge shapes. In matrix representations, the cells can be color-coded or be replaced by small icons to show edge attributes; node attributes might be shown as colored node labels, for instance. Usually, all mentioned efforts to integrate multivariate attributes into network representations do not scale well and get easily cluttered. The next subsection provides a more detailed classification of techniques which go beyond the traditional graph drawing approaches.

Classification of Approaches

Good drawing algorithms as previously described cannot solely solve the problem of MVNs. There are several reasons for this statement. First, the most traditional graph drawings do not scale well, i.e., they are not able to represent huge data sets with many thousands of nodes and/or edges. Second, additional multivariate data cannot be intuitively embedded into a standard drawing. The InfoVis community has tried to address those issues by visualization approaches that provide filtering and interaction possibilities in order to reduce the number of graph elements under consideration as well as by methods to visually analyze attributes in context of the underlying graph topology. According to Jusufi [12], several approaches can be found in the literature that offer solutions for the problem of visualizing multivariate networks: *multiple and coordinated views*, *integrated approaches*, *semantic substrates*, *attribute-driven layouts*, and *hybrid approaches*.

Multiple and coordinated views: Solutions in this category combine several views and present them together. This strategy allows the user to choose the most powerful visualization techniques for each specific view and data set [9, 25]. As an application example, we highlight the work of Shannon et al. [27]. Their approach consists of two distinct views: one view shows a parallel coordinate approach for the visual representation of the network attributes, and the other view displays a traditional node-link drawing of a graph. The tool is equipped with a variety of visualization and interaction techniques; both views are coordinated by linking and brushing [29] techniques. The drawback of multiple views is that they split the displayed data because of the spatial separation of the visual elements.

Integrated approaches: To provide a combined picture, attributes and the underlying graph can be displayed in one single view. "Integrated views can save space on a display and may decrease the time a user needs to find out relations; all data is displayed in one place." [9]. In Borisjuk et al. [2], small diagrams (e.g., bar charts) are employed instead of representing the nodes as simple circles, dots, or rectangles. Each diagram shows experimental data that is related to the regarded node. This approach provides a view of all available information, but the embedding of the visualizations into the nodes consumes a lot of space. This issue may affect the readability of the network due to the visual clutter that may appear when the number of nodes and the attributes is high [17]. However, the problem of space usage and additional clutter can be alleviated by interaction techniques.

Semantic substrates: In order to further avoid clutter in multivariate network visualizations, some researchers realized the idea of so-called semantic substrates that "are non-overlapping regions in which node placement is based on node attributes": Shneiderman and Aris [28] introduced this idea and combined it with sliders to control the edge visibility and thus to ensure comprehensibility of the edges' end nodes. Their tool efficiently improves the situation of visual clutter that happens with large MVNs. However, one conceptual drawback of such approaches is that the underlying graph topology is not (completely) visible.

Attribute-driven layouts: Those layouts use the display of the network elements to present insight about the attached multivariate data instead of visualizing the graph topology itself. In contrast to semantic substrates, this technique does not necessarily place the nodes into specific regions. Instead, it controls the placement of a node in the graph layout by considering the node's attributes. An example is *PivotGraph* [33] which shows the relationships between (node) attributes and links within a 2D grid-layout. This concrete approach scales well for some situations because of the inherent node aggregation (nodes on the same grid position share the same attribute values) but is restricted to discrete attribute values and only two attribute dimensions.

Hybrid approaches: They combine at least two of the previously discussed techniques. The most common combinations are multiple coordinated views with any of the integrated approaches. For instance, Rohrschneider et al. [26] integrate additional attributes of a biological network inside the nodes and edges. The authors also use other visual metaphors for creating multiple coordinated views to show time-related data of the network. Another hybrid approach is the JauntyNets tool [13] which combines multiple coordinated views with an attribute-driven layout.

1.3 Outline of This Book

The book is divided into two parts. The first three chapters (Chaps. 2-4) present three application domains in which multivariate networks are commonly used: software engineering, social networks and the life sciences. Written by experts in the three respective fields, these chapters describe how multivariate networks play a crucial role in the study of the comprehension of programs for the purposes of maintenance and evolution (Chap. 2), the analysis of personal and social networks defined by a wide variety of relationships (Chap. 3), and the exploration and analysis of biological data at several levels of detail (Chap. 4). Not only do these chapters describe the use of multivariate networks in these domains, they also consider how these networks can be effectively and appropriately visualized so as to support domain-specific tasks, and discuss the challenges facing these three rapidly evolving fields.

The second part of the book covers a range of topics associated with the visualization and use of multivariate networks, focussing first on fundamental visualization aspects (tasks, interaction, and representation), and then addressing broader issues (time, multiple networks, and large networks). Chapter 5 presents a new framework of tasks specifically associated with multivariate networks, based on existing taxonomies of general visualization tasks and simple graph-reading tasks. These multivariate network tasks are shown to be composed of lower-level visualization tasks, and are then illustrated with domain-specific examples. Chapter 6 highlights the fact that effective completion of user tasks when using a visual representation requires interaction, allowing the information landscape to be navigated, and more of the information to be perceived. It describes the range of different methods of interacting with multivariate networks, as well as guidelines for novel interaction techniques. Chapter 7 focuses on the means by which multivariate networks can be visually represented—beyond the traditional node-link method—by proposing and discussing a range of alternative (and novel) visual metaphors inspired by nature, geography or manufactured objects. It concludes with a gallery of potential new metaphors.

The important issue of time is covered in Chap. 8. This chapter provides essential definitions for temporal multivariate networks, and shows how two applications (biology and social networks) relate to a structure-behaviour-evolution model originally proposed for characterizing temporal networks in software engineering. A survey of existing visualization methods for temporal networks is presented. The heterogeneous networks chapter (Chap. 9) is primarily concerned with multivariate networks that are associated with each other at different levels and at different scales, and demonstrates the concepts with examples from the three application domains of biology, social sciences, and software engineering. The challenges of visualizing such linked networks are discussed. The final chapter (Chap. 10) considers the ever-present visualization challenge of scalability—what to do when the networks are so large that they cannot be displayed effectively. Based on considerations of cognitive and architectural limitations, suitable visualization approaches for large networked data sets are explored, and their effectiveness discussed.

References

1. Abello, J., van Ham, F.: Matrix zoom: A visual interface to semi-external graphs. In: Proceedings of the IEEE Symposium on Information Visualization, pp. 183–190. IEEE Computer Society, Los Alamitos (2004)
2. Borisjuk, L., Hajirezaei, M.R., Klukas, C., Rolletschek, H., Schreiber, F.: Integrating data from biological experiments into metabolic networks with the dbe information system. In Silico Biol. 5(2), 93–102 (2005)
3. Chambers, J.M., Cleveland, W.S., Kleiner, B., Tukey, P.A.: Graphical Methods for Data Analysis. Wadsworth, Belmont (1983)
4. Chaomei, C.: Information Visualization. Beyond the Horizon, 2nd edn. Springer, Heidelberg (2004)
5. Chernoff, H.: The use of faces to represent points in k-dimensional space graphically. Journal of the American Statistical Association 68, 361–368 (1973)
6. Cleveland, W.C., McGill, M.E.: Dynamic Graphics for Statistics. CRC Press, Inc., Boca Raton (1988)
7. Battista, G.D., Eades, P., Tamassia, R., Tollis, I.G.: Graph Drawing: Algorithms for the Visualization of Graphs. Prentice Hall (1999)
8. Dwyer, T., Hong, S.H., Koschützki, D., Schreiber, F., Xu, K.: Visual analysis of network centralities. In: Misue, K., Sugiyama, K., Tanaka, J. (eds.) Proceedings of the 2006 Asia-Pacific Symposium on Information Visualisation (APVis 2006). ACM International Conference Proceeding Series, pp. 189–198. Australian Computer Society, Darlinghurst (2006)
9. Görg, C., Pohl, M., Qeli, E., Xu, K.: Visual Representations. In: Kerren et al. [17], pp. 163–230
10. Inselberg, A., Dimsdale, B.: Parallel coordinates: A tool for visualizing multi-dimensional geometry. In: IEEE Visualization, pp. 361–378 (1990)
11. Johnson, B., Shneiderman, B.: Tree-maps: a space-filling approach to the visualization of hierarchical information structures. In: Proceedings of the 2nd Conference on Visualization 1991, VIS 1991, pp. 284–291. IEEE Computer Society Press, Los Alamitos (1991)
12. Jusufi, I.: Multivariate Networks: Visualization and Interaction Techniques. Ph.D. Thesis, Linnaeus University, Växjö, Sweden (2013)

13. Jusufi, I., Kerren, A., Zimmer, B.: Multivariate Network Exploration with JauntyNets. In: Proceedings of the 17th International Conference on Information Visualisation (IV 2013), pp. 19–27. IEEE Computer Society Press, Los Alamitos (2013)
14. Kaufmann, M., Wagner, D. (eds.): Drawing Graphs. LNCS, vol. 2025. Springer, Heidelberg (2001)
15. Keim, D.A.: Information visualization and visual data mining. IEEE Transactions on Visualization and Computer Graphics 7(1), 1–8 (2002)
16. Keim, D.A., Hao, M.C., Dayal, U., Hsu, M.: Pixel bar charts: A visualization technique for very large multi-attribute data sets? Information Visualization 1(1), 20–34 (2002)
17. Kerren, A., Ebert, A., Meyer, J. (eds.): Human-Centered Visualization Environments. LNCS, vol. 4417. Springer, Heidelberg (2007)
18. Kerren, A., Schreiber, F.: Network visualization for integrative bioinformatics. In: Chen, M., Hofestädt, R. (eds.) Approaches in Integrative Bioinformatics, pp. 173–202. Springer, Heidelberg (2014)
19. Kerren, A., Stasko, J.T., Fekete, J.-D., North, C. (eds.): Information Visualization. LNCS, vol. 4950. Springer, Heidelberg (2008)
20. Koschützki, D., Schreiber, F.: Comparison of centralities for biological networks. In: R. Giegerich, J.S. (ed.) Proc. German Conf. Bioinformatics (GCB 2004). pp. 199–206 (2004)
21. Mardia, K.V.: Multivariate Analysis. Academic Press (1979)
22. Newman, M.E.J.: Networks: An Introduction. Oxford University Press (2010)
23. Novak, J.D.: Learning, Creating, and Using Knowledge: Concept Maps as Facilitative Tools in Schools and Corporations, 2nd edn. Routledge (2010)
24. Pickett, R.M., Grinstein, G.G.: Iconographic displays for visualizing multidimensional data. In: Proceedings of the 1988 IEEE International Conference on Systems, Man, and Cybernetics, vol. 1, pp. 514–519 (1988)
25. Roberts, J.C.: Exploratory visualization with multiple linked views. In: MacEachren, A., Kraak, M.-J., Dykes, J. (eds.) Exploring Geovisualization, ch. 8, pp. 159–180. Elseviers (2004)
26. Rohrschneider, M., Ullrich, A., Kerren, A., Stadler, P.F., Scheuermann, G.: Visual network analysis of dynamic metabolic pathways. In: Bebis, G., et al. (eds.) ISVC 2010, Part I. LNCS, vol. 6453, pp. 316–327. Springer, Heidelberg (2010)
27. Shannon, R., Holland, T., Quigley, A.: Multivariate graph drawing using parallel coordinate visualisations. Tech. Rep. 2008-6, University College Dublin, School of Computer Science and Informatics (2008), http://www.csi.ucd.ie/files/ucd-csi-2008-6.pdf
28. Shneiderman, B., Aris, A.: Network visualization by semantic substrates. IEEE Transactions on Visualization and Computer Graphics 12, 733–740 (2006)
29. Spence, R.: Information Visualization: Design for Interaction, 2nd edn. Prentice Hall (2007)
30. Stasko, J.T., Catrambone, R., Guzdial, M., Mcdonald, K.: An evaluation of space-filling information visualizations for depicting hierarchical structures. International Journal of Human-Computer Studies 53, 663–694 (2000)
31. Ward, M., Grinstein, G., Keim, D.A.: Interactive Data Visualization: Foundations, Techniques, and Application. A.K. Peters, Ltd. (2010)
32. Ware, C.: Information Visualization: Perception for Design, 2nd edn. Morgan Kaufmann (2004)

33. Wattenberg, M.: Visual exploration of multivariate graphs. In: CHI 2006: Proceedings of the SIGCHI Conference on Human Factors in Computing Systems, pp. 811–819. ACM, New York (2006)
34. Williams, M., Munzner, T.: Steerable, progressive multidimensional scaling. In: Proceedings of the IEEE Symposium on Information Visualization (InfoVis 2004), pp. 57–64. IEEE Computer Society Press (2004)

Part I

Application Domains – Characteristics and Challenges

2
Multivariate Networks in Software Engineering

Stephan Diehl and Alexandru C. Telea

Multivariate networks, or graphs, occur in many application domains. In this chapter, we focus on software engineering. We present the specific nature of the data, challenges, and visual exploration solutions for multivariate graphs stemming from software engineering applications. Our goal is twofold. First, we draw attention to specific software engineering aspects, and the ensuing multivariate graphs, which make their (visual) understanding hard. This should help researchers to better understand the software engineering challenges related to multivariate graphs, and thus contribute to solutions. Secondly, we present existing approaches for the visual exploration of multivariate software graphs. This should help disseminating such solutions to areas beyond software engineering.

The structure of this chapter is as follows. In Sect. 2.1, we outline the importance and scope of software visualization. Section 2.2 details the characteristics of the data involved in such visualizations and the scope of multivariate graphs herein. Section 2.3 presents a selection of relevant tasks addressed by software visualization which involves multivariate graphs, and also presents visualizations that address such tasks. Section 2.4 discusses the current state-of-the-art in multivariate visualization of software networks, and outlines the main challenges that this application domain currently faces.

2.1 Aims and Scope

To understand the specific challenges (and existing solutions) to the visualization of multivariate software graphs, we first need to understand the main aims and scope of software visualization (SoftVis). In this section, we provide an overview answer to this question. Given the huge scope of software engineering and, implicitly, SoftVis techniques and tools, we cannot aim at a complete review. Rather, the aim is to outline the key value drivers that make SoftVis relevant to software engineering, and also to highlight how software visualization (with a focus on multivariate graphs) differs from other multivariate graph visualizations. For a comprehensive survey of software visualization, we refer to [12].

2.1.1 History and Definitions

Early examples of software visualizations include the visual depiction of program control flow charts [16, 36], sorting algorithms [2], and software source code [13]. In the 1990s, software visualization was being recognized as a separate research field. One of its first definitions is as follows: "Software visualization is a representation of computer *programs*, associated *documentation* and *data*, that enhances, simplifies and clarifies the *mental representation* the software engineer has of the operation of a computer system" [39]. We see that SoftVis covers the full range of data artifacts produced by the software lifecycle. Equally importantly, we see that the key aim of SoftVis is to help software engineers to *understand* the operation of software systems. These aspects have stayed relevant throughout the history of software visualization, as further discussed.

A decade later (2007), a comprehensive survey of software visualization [12] proposed the following definition: "Software visualization targets the visual depiction of the structure, behavior, and evolution of software". The definitions of these three key data ingredients of software are as follows:

1. **Structure:** describes all entities involved in the studied software, including their properties and relations between them;
2. **Behavior:** describes how entities dynamically interact with each other, and also process data, during program execution;
3. **Evolution:** describes how software is changed during the software lifecycle.

The reach of software visualization to software evolution parallels the growing interest in developing models, techniques, and tools for the data mining and analysis of software evolution processes [26]. In parallel, the audience of SoftVis is also enlarged, to include almost all stakeholders of the software lifecycle: product and process managers, architects, designers, developers, and testers [23].

2.1.2 Importance

To better advocate the necessity and added value of SoftVis, we consider the question: Is software visualization really needed? Answering this question has two parts. First, its application domain, the software industry, is large and growing: $457 billion for 2013, 50% larger than in 2008 [21]. For comparison, the total US healthcare spending in 2009 was $2.5 trillion [47]. Studies over two decades show that 80% of software development costs are spent on maintenance [10, 26, 38]. Secondly, over the same period of time, several studies have shown that over 50% of the effort spent by software engineers is dedicated to *understanding* the software [12, 23, 26]. As modern software systems become even larger, this understanding effort becomes a key component of the software lifecycle [5].

2.2 Data Characteristics

Data involved in program comprehension is large, complex, and changes in time. As such, software visualization has a good potential to be an effective part of comprehension solutions. In a survey of over 100 practitioners involved in software maintenance and re-engineering, 42% of the participants stated that SoftVis is an important, but not critical, aid to comprehension; 42% other participants found SoftVis absolutely necessary for their work [23]. A survey on SoftVis tools highlighted as added value points the increase of productivity and quality of produced software, better management of complexity in large software systems, all leading to saving time and money in development and maintenance [3].

Software can be modeled by three orthogonal data aspects: *entities*, *relations*, and *attributes*. These are detailed in the following sections.

2.2.1 Entities

Software entities correspond to the *nouns* in the software description, i.e., describe the items which interact to form the structure, behavior, and evolution of a software system. Structural entities typically describe the static organization of a software corpus. Examples are folders, files, packages, components, classes, methods, and individual lines of source code. Behavioral entities describe the execution of a software system. Examples are program traces, profiling logs, method invocations, test results, and bug reports. Evolutionary entities are, largely speaking, related to the process of software maintenance. Examples are change requests, product documentation and requirement documents, development tasks, and actors with different roles in the development process (contributors to software repositories, testers, quality engineers, and release managers).

2.2.2 Relations

Relations correspond to the *verbs* in the software decription, i.e., describe how various software entities are connected and interact to form the structure, behavior, and evolution of a software system. Structural relations can be further organized into *hierarchical* and *association* relations.

Hierarchical relations describe the static structure of a software system. They form a part-whole hierarchy that captures the aggregation of smaller-scale software entities into larger units. Examples of containment relations are

1. **physical** relations (files in folders in higher-level folders);
2. **logical** relations (methods in classes in libraries in systems).

Several such hierarchies may be needed to describe a given system. For instance, C++ programs admit both a physical file-folder hierarchy and a logical namespace-class-method hierarchy, and the two hierarchies are not identical.

Association relations cover all relations which do not describe (hierarchical) part-whole relationships. Examples are

- **calls:** function A calls function B;
- **inheritance:** class A inherits from class B;
- **co-change:** file f_1 changed at the same time as file f_2;
- **duplication:** files f_1 and f_2 share similar (cloned) source code;
- **change impact:** when changing file f_1, we next need to change file f_2;
- **ownership:** developer D performed task T; class X owns an object Y;
- **data flow:** component C reads data from component Y.

Clearly, several types of relations are needed to describe the structure, behavior, and evolution of a software system. Association relations can form both acyclic graphs (e.g., inheritance) but also cyclic graphs (e.g., a call stack containing recursive or re-entrant functions). Associations can be either undirected (e.g., clone or co-change relations) or directed (e.g., inheritance or call). Software associations are typically one-to-many relations (e.g., a function calls several other functions; a data object owns a collection of subordinate objects).

2.2.3 Attributes

Attributes model structural, behavioral, and evolutionary *properties* of both entities and relations. Examples are

- **syntax:** name, signature, and location in the source code for classes, functions, or individual symbols;
- **execution:** call duration, call stack depth, processor allocation, and resource usage of a function call in a program trace;
- **person:** name, role, and e-mail of a person involved in a software maintenance process;
- **testing:** time stamp, number of failed and passed tests, and amount of lines of code covered by tests for a code unit.

Typically, attributes are modeled as key-value pairs for an entity or relation, e.g., a class C has an attribute *name* with value C. An entity or relation can have several such attributes. Also, entities (or relations) of the same kind, or type, do not necessarily need to have the same number of attributes. This is, among others, due to incomplete data delivered by the various data mining tools used.

2.2.4 Software as Multivariate Time-Dependent Graphs

Entities, relations, and attributes can change during the lifetime of a software product. Two causes drive this process:

1. **behavior:** running the same software system several times can yield different execution paths and data values. Thus, both relations (calls) and attributes (call durations) for the same entity (caller function) will differ;
2. **evolution:** software continuously changes as it is maintained. Hence, an entity (e.g., file) can have different contents, relations (to other files), and developers owning it over time.

Putting it all together, we can describe software as a *multivariate time-dependent graph* $G = (V, E = V \times V)$. Nodes V model software entities. Edges E model structure and association relations. Each node $n \in V$ and edge $e \in E$ has a set of attributes $\{a_i^n\}$ and $\{a_i^e\}$ respectively. Each attribute $a = (key, val)$ is a key-value pair. Keys key are typically textual or categorical identifiers. Attribute values val belong to various domains, depending on the attribute type, e.g. $\mathbb{N}$ for code size, $\mathbb{R}^+$ for execution duration, $\mathbb{B}$ for test outcomes (passed, failed), or Σ^* for developer names (where Σ is the used alphabet). All elements of G, i.e., V, E, a_i^n, and a_i^e are time-dependent, i.e., functions of $t \in T$. Since both software execution and software evolution are discrete processes, and also since data is mined from software systems typically at discrete points in time, T is usually a finite set of ordered points in time $T = \{t_i \in \mathbb{R}^+ | t_i < t_j, \forall i < j\}$.

2.2.5 Reference Implementation

Hierarchy-and-association graphs G are also often called *compound* graphs in the literature [41]. Creating, storing, manipulating, and ultimately understanding the information captured by such graphs is clearly very challenging, even for moderately-sized systems. Important questions in this respect are

1. **schema:** How to best model (capture) a given aspect of a software system in terms of entities, relations, and attributes?
2. **selection:** How to select data relevant for a given task from an entire G?
3. **implementation:** How to store G in a way that is efficient for quickly reading and writing large amounts of data?

Several so-called *data schemes* or data models for G have been proposed, e.g., [14, 25, 28, 31, 44]. This wealth of models can be puzzling for the practitioner interested in using existing SoftVis tools and techniques on given software datasets. More importantly, not all challenges of modeling multivariate software data become evident from studying such data schemes.

To outline such challenges, we present next a data model for G. This model is based on a SQL relational database, so it is simple to understand, scalable, computationally efficient, and can generically capture most degrees of freedom of G. First, we define the concept of a *selection*. Selections $S \subset G$ are subsets of nodes and/or relations that specify the part of G that we want to study. They are a necessary abstraction when using a single graph G to store all available data, or facts, mined from a software project. For example, if we are interested in studying the call graph of a system, we need a selection containing only software-structure nodes and call relations between them. If we want to study the contribution of a given developer, we need a selection containing only entities and relations that the respective developer has worked on.

The proposed SQL schema contains the following elements (see also Fig. 2.1 top):

1. **Keys:** each node, hierarchical and association edge, and selection, has a unique ID (further used as primary key);
2. **Hierarchy** table: one hierarchical edge, listed as $(parent, child)$ node IDs, per row;
3. **Association** table: an association edge, listed as $(from, to)$ node IDs, per row;
4. **Node attribute** table: each row stores all attributes (metrics) $a_1^n, \ldots, a_k^n$ of a given node n as k columns;
5. **Association attribute** table: each row stores all attributes (metrics) $a_1^e, \ldots, a_k^e$ of a given association edge e as k columns; different edge types, *e.g.* calls, uses, includes, are modeled by adding an edge-type attribute;
6. Two **selection** tables per selection $s \in S$, for the node IDs and edge IDs of the items in s, respectively.

Figure 2.1 (bottom) illustrates this schema this for a simple program. Hierarchy consists of a file *main.cc* containing two functions *main()* and *run(Foo)*, and a class *Foo* with a method *load()*. Associations are call, define, and 'uses type' relations, modeled as edge 'type' attributes. Nodes have two attributes: name and lines-of-code size (LOC). Two selections exist: the call graph of *main()* (red), and the 'uses' graph of *run(Foo)* (green).

This schema can store any compound (hierarchy-and-association) attributed graph, e.g., annotated syntax graphs, call graphs, developer networks, or code duplication relations. New association types can be added to a database without changing its schema, since types are stored as attributes. This allows incrementally refining an existing *fact database*, e.g., by adding new results obtained from additional data mining processes. Adding node or relation attributes amounts to adding new columns for the node and edge attribute tables respectively. Attribute types can be any of the supported data types of the underlying SQL database (e.g., numeric, text, date-time, image, or binary blob). Multiple association types can be stored in a single pair of association and association-attribute tables. Hierarchy data is stored separately in a hierarchy table. This follows the observation that, in software databases, hierarchy data is much smaller in size (and typically changes less frequently) than association data. As such, this schema is more efficient for

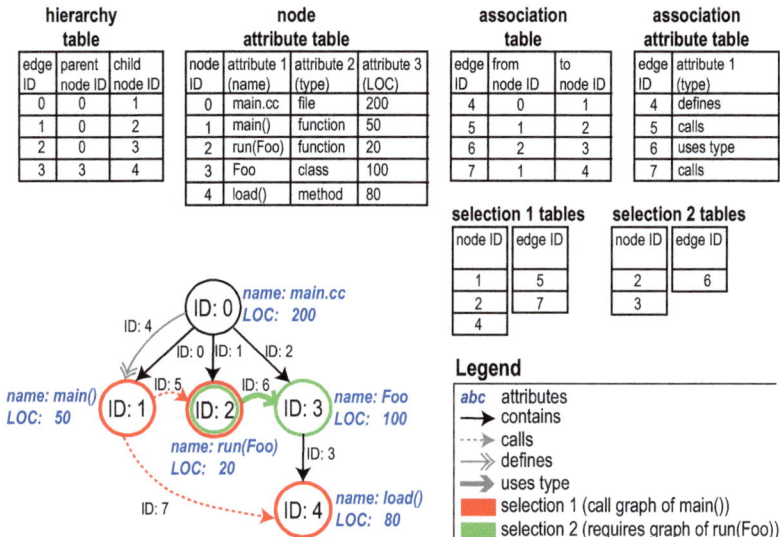

Fig. 2.1. Database schema (top) for a compound attributed graph (bottom) and two selections: The call graph of $main()$ and the 'uses' graph of $run(Foo)$

fast querying and updating. If desired, several hierarchy tables can be used to model multiple software hierarchies (Sect. 2.2.2).

Multiple selections can be stored in separate selection tables. This fully decouples data *storage* (node, association, and hierarchy tables) from data *usage* (e.g., visualization). Users can create and iteratively refine selections by executing SQL queries on already existing selections. This supports the visual information-seeking mantra "overview, filter, then details on demand" [37]. The above schema scales well to databases of millions of entities and relations [34].

However, the above schema for G does not capture time-dependency. Schemas that model time-dependent graphs have been proposed, e.g., [31, 49]. However, to be scalable to large software projects, such schemas are geared towards capturing specific types of relations, rather than the generic model of the graph G outlined above. A fully general solution to efficiently and effectively modeling G is not yet known, and this is a topic for future research.

2.2.6 Software Data *vs.* other InfoVis Domains

Large multivariate time-dependent graphs having the model outlined in Sect. 2.2.5 are not unique to software engineering. They arise also in other application areas, most notably biology, chemistry, and bioinformatics. As such, relevant questions are: What is specific to the software understanding domain, which underlies the evolution of software visualization as a discipline separated from biological visualization (BioVis) or, more generally, information visualization (InfoVis)?

BioVis: Comparing our software graphs to networks in the BioVis domain, we notice several similarities: In both domains, large graphs (hundreds of thousands of entities, relations, and attributes or more) are common in real-world use-cases. As such, scalability and efficiency are common concerns. Moreover, both domains feature problems involving multivariate and time-dependent graphs. However, several differences exist. First, SoftVis artifacts, and thus graphs extracted from them, are *man made*. In contrast, BioVis graphs capture (the measurement of) natural processes. In other words, SoftVis graphs are *constructive*, whereas BioVis graphs are *observational*. This deserves additional explanations. We could, on the one hand, say that SoftVis graphs are also observational, if we consider a software process as a "black box" which is monitored from the outside, e.g., to reverse-engineer its behavior. On the other hand, software *is* constructed by humans. As such, the underlying software understanding process takes the form of recovering (possibly lost) semantics. In contrast, understanding biological processes often aims at discovering yet-unknown natural laws and designs.

A second difference relates to uncertainty. Software processes are defined by an underlying *exact* computational model given by the processor and semantics of used programming languages. For example, there is no uncertainty as to which is the type-usage or inheritance graph of a given code base. In contrast, BioVis data typically contains more uncertainty, due to measurements and the natural variability of experiments. If, however, we add human aspects to SoftVis data, e.g., we want to reason about developer properties, then data uncertainty becomes more important, and this distinction gets blurred.

Last but not least, a major distinction is induced by the *user group*. In BioVis, users are typically not computer scientists themselves. As such, they are likely less familiar with various algorithmic, implementation-level, and data modeling aspects involved in the construction and usage of visualization tools. In contrast, developers and users of SoftVis tools largely overlap – they are all computer science professionals. On the one hand, this makes the task of SoftVis developers easier, as they understand both the end goals and mechanisms their tools should support and respectively provide. In contrast, developing effective BioVis tools is a much harder proposition, as their developers have to become, at some point, experts in *both* information visualization and biology.

InfoVis: Software visualization can be seen as a specialized sub-branch of information visualization (InfoVis). However, if we compare the focus of many InfoVis research projects with their SoftVis counterparts, several differences emerge. First and foremost, SoftVis 'solutions' (techniques, tools, and applications) show a strong coupling of data mining and visualization components, covering the entire pipeline from getting the raw data, filtering and analyzing this data to extract relevant information, and next exploring this information visually to (in)validate a hypothesis related to a software process or product. As such, SoftVis is closer to what is currently called visual analytics. In contrast, a significant part of InfoVis research focuses on more generic problems, such as the visualization of large *generic* graphs, tables, or hierarchies. Typical program understanding challenges involve correlating a multitude of different aspects, such as source code, execution traces, documentation, and developer activities. As such, SoftVis datasets are by nature high-variate graphs which contain attributes of a multitude of different *types*. The challenge of visually understanding multivariate data is thus fundamental to SoftVis. In contrast, multivariate data is not, by definition, a key aspect to *all* InfoVis applications and solutions.

2.3 Applications

In the previous section, we have shown that SoftVis datasets consist naturally of large multivariate time-dependent attributed graphs. In this section, we overview a number of techniques and tools that have been developed in the SoftVis domain for visualizing such data. We organize the presentation along the structure, behavior, and evolution aspects introduced earlier. For each aspect and solution, we also outline the *tasks* that the respective solution aims to support (Fig. 2.2), and also emphasize the multivariate nature of the visualized data.

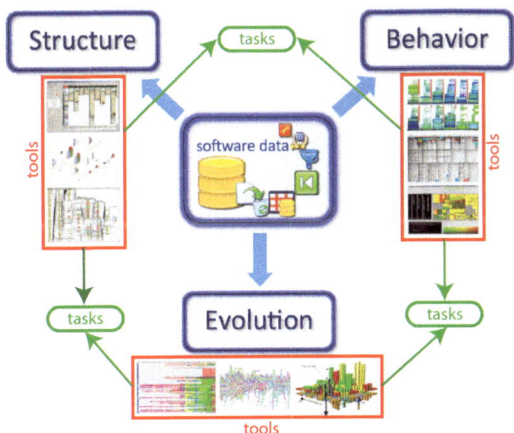

Fig. 2.2. Data sources, tools, and tasks in software visualization

2.3.1 Structure Visualization

Software has a hierarchical structure (Sect. 2.2.2). At the lowest level, we can visualize individual lines of code. Figure 2.3 shows two examples. Both examples share the same core idea, introduced by the SeeSoft tool [13]: show each code line as a horizontal pixel line, scaled by the line's length (in characters), and colored by a data attribute computed for that line. Similar to the table lens [32], line-level visualizations scale well up to tens of thousands of code lines on a single screen. Image (a) shows the Tarantula tool [22]. Here, lines are colored by a data value indicating testing outcomes. Red lines show many failures, green lines show passed tests, and gray lines show code not covered by tests. Image (b) shows a similar design in the CSV tool [24]. Here, colors are added to syntax blocks in source code, rather than individual lines. Users can pick specific language constructs, such as functions, class declarations, iterative statements, conditional statements, variables, or comments, using a classical tree browser for the language's syntax, and assign them specific colors. Matching code blocks are displayed using these colors by the shaded cushion technique introduced by van Wijk and van de Wetering for treemaps [48]. The spatial cushion nesting conveys the code's nesting depth. The color distribution conveys the overall code structure. For instance, in Fig. 2.3b, green shows comments. We can thus see that the visualized code (around 10K lines) is densely and uniformly commented.

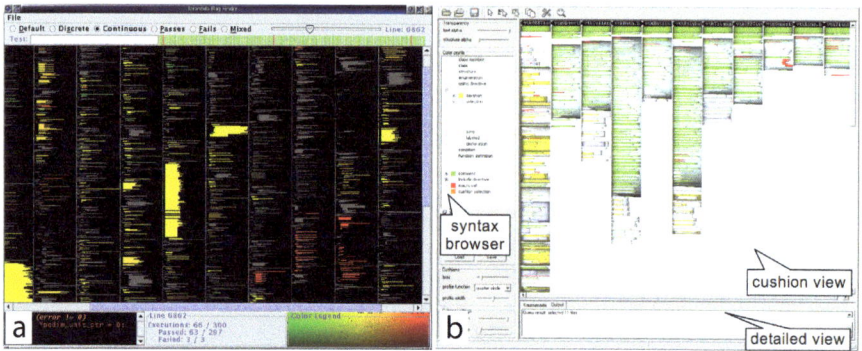

Fig. 2.3. Line-level (a) and syntax-level (b) visualization of program structure

Line-based visualizations have a natural multiscale aspect. Zooming out, we can continuously transition from the simplified images in Fig. 2.3 to classical text views where individual code lines are readable [24]. Several software aspects can be viewed simultaneously: structure-and-evolution (Fig. 2.3a), and structure-and-behavior (Fig. 2.3b). However, we also see several limitations. First, in the continuous transition described above, continuous *attribute interpolation* is hard to do for non-numerical attribute types. Secondly, given the limited display space, it is hard to show several attributes per item (line of code).

Structure at higher levels than code lines involves folders, files, classes, and methods (Sect. 2.2.2), and also their relations. Figure 2.4 shows several such visualizations. Image (a) shows a classical UML diagram (nodes=classes,

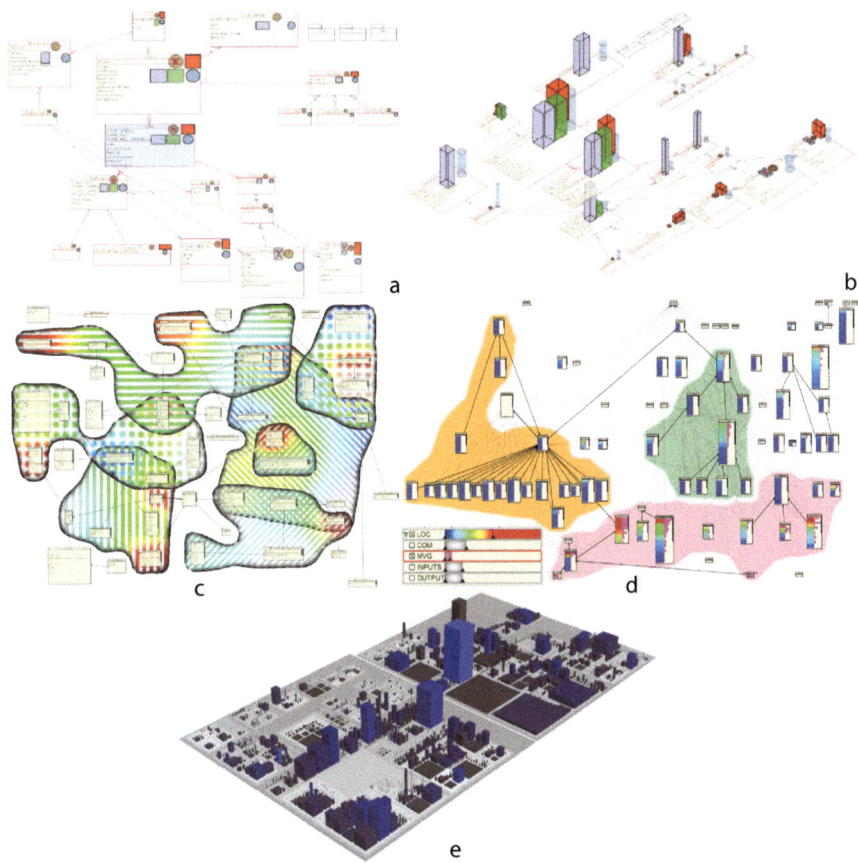

Fig. 2.4. Code structure and multiple code metrics shown on UML and treemap diagrams

edges=interitance, 'has' and 'uses' relations). Nodes are laid out using graph drawing algorithms such as Sugiyama-style methods or spring embedders [15]. Multiple attributes, e.g., code quality metrics, are shown atop of each class using glyphs (bar and pie charts) sized and colored by the metric values. Glyphs are laid out in the same (grid) order in each class. This helps correlating the same attribute across different classes. As typical UML diagrams contain only tens of nodes (classes), nodes offer enough space to show several per-class metrics. Image (b) shows how the third dimension brings an additional degree of freedom—here, glyph heights attract attention to extreme attribute values. This 3D technique is generalized in CodeCity [50], where the UML 'base' layout is replaced by a treemap to increase information density (Fig. 2.4e).

Additional structure can be added by considering so-called *areas of interest* (AOIs). AOIs are sets of nodes which share a common property, e.g., all thread-safe or all platform-dependent classes in a system [8]. Such sets can be nested, overlap, or be disjoint. AOIs can be shown with Venn-Euler diagrams, by surrounding all elements in a set by a smooth shape (Fig. 2.4c). Adding shaded cushions [48] helps seeing how AOIs overlap or nest. Nodes can also have *per-AOI* metrics. These are multiple attribute values that a node has, one for each AOI it belongs to—for example, the amount of thread-safe, respectively platform-dependent code lines that a class has. To visualize these attributes, space is used outside the node icons (which are reserved to show the AOI-independent node attributes)—specifically, per-AOI attributes are drawn on the AOI cushions using texture and color interpolation. Texturing creates a weaving pattern which helps mapping the identity of an attribute to its corresponding AOI.

Detail information can be added by a table lens [32] atop each class icon (Fig. 2.4d). Rows are class methods, and columns show 1..3 metrics for each method. All class tables can be sorted synchronously, which allows easily comparing the metrics' distributions across an entire diagram.

Despite considerable work in the graph drawing community, classical node-link diagrams are effective only for graphs up to a few hundred nodes. Beyond this, clutter created by node-node, edge-edge, and node-edge overlaps makes reading such images hard. Also, for large graphs, the white space left between nodes by such algorithms makes their use less scalable. A different approach is taken by hierarchical *edge bundling* (HEB) methods. Pioneered by Holten [18], HEB assumes its input is a compound (hierarchy-and-association) graph. Starting with a *given* node layout, HEB groups, or bundles, straight-line association edges between these nodes using the hierarchy relations. If the given node layout is compact (space-filling), then HEB can scale easily to thousands of nodes and edges on a single screen. Additional automatic level-of-detail and interaction options make HEB scale to hundreds of thousands of nodes and edges [20].

Figure 2.5 shows the SolidSX Software eXplorer tool [34]. The input compound graph captures the syntactic structure of a C# program (50K lines of code). Image (a) shows the program hierarchy (assemblies, classes, methods, and files) using a classical tree browser. In image (b), a table lens [32] shows several method-level attributes (name, size, complexity, number of callers, and number of callees). Sorting this table allows finding, e.g., the most complex and/or largest methods. In image (c), these methods are highlighted atop of the software structure-and-association graph shown with HEB. Relations (calls, inheritance, and type usage) between elements are shown by bundles, colored by relation type. Finally, a fourth view, image (d), uses a treemap to show two different node attributes encoded in the treemap-cell colors and sorting order.

Several aspects are relevant here. Visual scalability is achieved by using different types of space-filling techniques: table lenses, treemaps, and HEB plots. Understanding aspects which are encoded in the correlation of *multiple* attributes is done by using multiple views linked by selection and brushing.

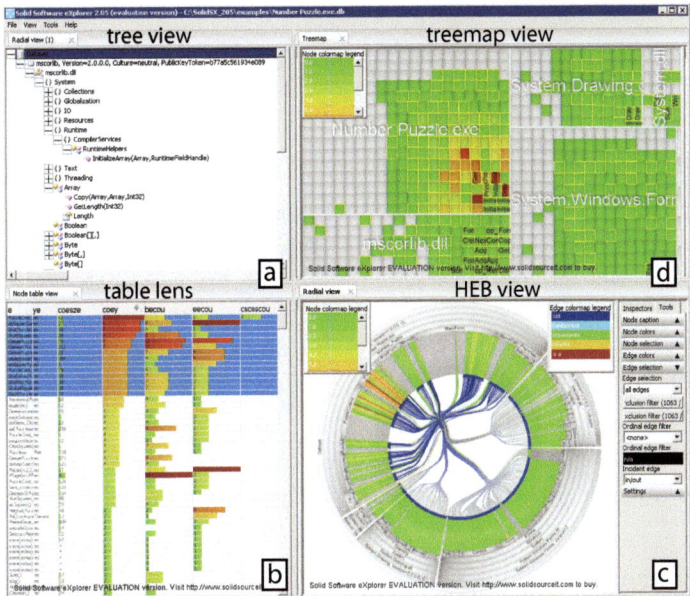

Fig. 2.5. Syntactic structure and attributes visualized with multiple space-filling views

This implicitly makes the entire solution scalable to multivariate data—the four-view display in Fig. 2.5 can show, in practice, ten such attributes per data element. However, this puts an extra burden on users in terms of performing the interactive view linking. Separately, bundling creates (by construction) many edge overlaps. Although this reduces visual clutter as compared to classical node-link displays, it also makes it hard to use color-mapping to show individual attributes at edge level. Also, even for limited amounts of edge overlaps, showing *multiple* attributes per edge in the same time is not possible.

Figure 2.6 shows a different use-case—the comparison of two hierarchies of a software system [4]. The choice of structures used supports different use-cases, e.g., the comparison of the logical and physical views (Sect. 2.2.2) of a system, or the comparison of two related systems such as two versions of a software code base. The horizontal and vertical icicle plots show the two considered structures. The shaded cells in the central adjacency matrix indicate how entities match between the two structures. In the shown figure, these cells are quite close to the diagonal, indicating a strong similarity of the two structures. Adjacency matrix plots are a good space-filling alternative to HEB views for showing compound graphs, and have been used to visualize very large call graphs [1, 17].

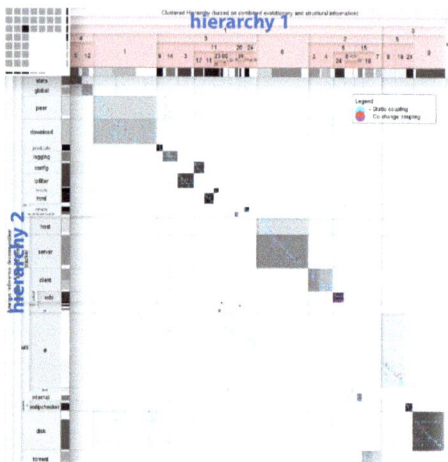

Fig. 2.6. Adjacency matrix visualization for comparing two software hierarchies

2.3.2 Behavior Visualization

Apart from program structure, behavior is an essential part of program comprehension. Captured by execution traces, behavior can be visualized using activity charts. Figure 2.7a shows a typical chart, produced by the Shark profiling tool on Mac OS X. Table rows correspond to function calls, sorted on calling order, call duration, or other user-specified criteria. The right table part shows per-CPU-core occupancy, color-coded by CPU code ID. This gives insight in how well a parallel program is designed to take advantage of a multi-threaded architecture. Similar techniques are used to visualize software behavior on superscalar-processors (Rivet tool [40]) and Java program executions (Jinsight tool [30]). Apart from the per-function-call view, the call stack metaphor is also used to visualize execution traces (Fig. 2.7b). Here, an icicle plot shows function call nesting and call duration. In this view, the call stack depth, as well as time spent in a function call itself vs. time spent in deeper-called functions, are easily visible.

Program structure can be added to execution trace data. The Extravis tool [11] does this by showing execution traces with a sequence view, where each call is drawn as a horizontal line (Fig. 2.7c, right). Line endpoints are aligned to match the layout of an icicle plot that shows the program static structure (Fig. 2.7c, top-right). This shows when, and how often, a given function declaration (in the static structure) was called during the execution. Separately, a HEB view shows the static structure and calls within a user-selected time range.

An alternative structure-and-behavior combination is proposed by the ViewFusion tool [46]. An icicle plot (Fig. 2.7d, top) shows the call stack, similar to Fig. 2.7b. This plot is overlaid atop of a treemap showing the

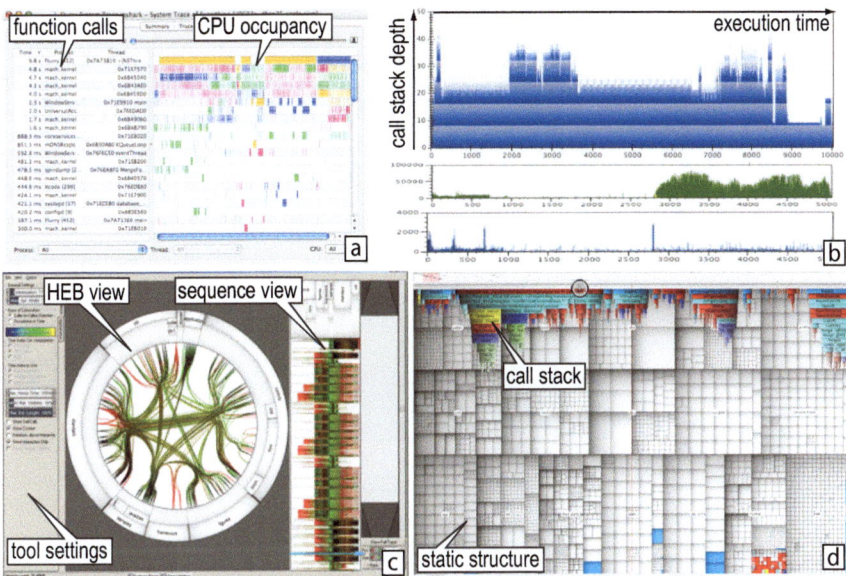

Fig. 2.7. Visualizations of execution traces (a-b) combined with program structure (c-d)

static system structure, and can be interactively panned and zoomed to select interesting execution time ranges. Function calls (from the call stack) are correlated with function declarations (from the treemap) using interaction and color mapping. Just as in Extravis, interaction and multiple views help cope with the multivariate data implied by program structure and execution information.

TraceDiff [45] extends the ViewFusion idea to compare two execution traces T_1 and T_2. (Fig. 2.8). The traces are shown at the top and bottom of the view, using the same icicle plot design as in Fig. 2.7d. HEB-like bundles are computed between function call sequences in T_1 and T_2 that match a user-supplied similarity criterion. To simplify the view, matching calls that are close to each other in both time and caller space are aggregated and rendered as thick shaded tubes, following a visual simplification technique originally proposed for HEB views [43]. Tubes are colored to encode the call similarity. Zooming the view in and out allows users to find matching call sequences at different levels of detail.

Multiple behaviors can also be visualized against software structure. The Gammatella tool [29] collects deployment data of multiple instances of a given software system, and shows a *distribution* of this data at detailed code-line level (Fig. 2.9, code view), SeeSoft-like level (file view), and package level (treemap structure view). On each shown element, the computed metric distribution is visualized using a color gradient ranging from red (failed) to green (successful).

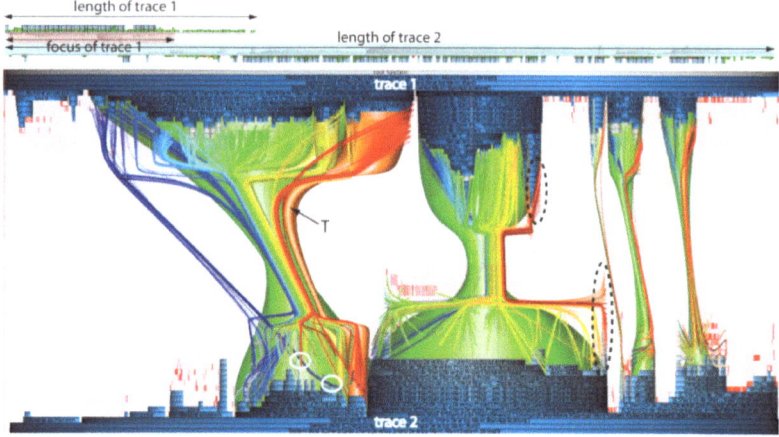

Fig. 2.8. Multiscale visual comparison of two execution traces

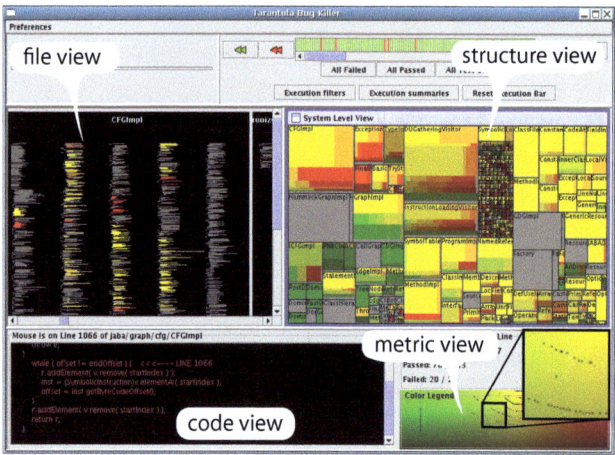

Fig. 2.9. Multiple deployment results *vs* system structure at different levels of detail

2.3.3 Evolution Visualization

Software evolution generates time-dependent data in terms of different versions, or revisions, of a software system. These can be mined from source control management (SCM) systems, or *repositories*, such as CVS, SVN, Git, or TFS. Each revision yields a multivariate compound attributed graph that can be visualized using the techniques described in Sects. 2.3.1 and 2.3.2. However, the sheer amount of data a typical repository stores is huge: thousands of files having tens of thousands of revisions spread over years. Repositories store additional data besides software structure and behavior, such as commit logs, change requests, time stamps, and the identity of developers who changed the software. This only increases the number of attributes available per data item. One approach is to reduce the amount of information by using rule or pattern mining techniques first. It turns out that the number of mined rules or patterns is still very large and that standard visualization techniques can be applied to interactively explore these rules [7]. Analyzing the data without the information loss induced by rule or pattern mining asks for different, more scalable, visualization techniques.

Figure 2.10a shows a first solution for evolution visualization [49], applied to a SVN repository. The x axis maps time. Each file is drawn as a horizontal line starting when the file was first committed in the repository. Lines are cut into chunks, one per interval between consecutive revisions. Chunk colors show an attribute, e.g., revision author, testing results, or code quality metrics. Files can be sorted along the y axis to support several analyses. In Fig. 2.10a, files are sorted by decreasing activity (revision density per unit time). This allows finding the most active files (placed at the top), and correlating these with other attributes, such as age, developer identity, or code metrics. In Fig. 2.10a, revisions are colored by developer ID. We see a large purple spot over the first evolution half for the top files, and a large green spot over the second evolution half. This shows that, halfway the project, the main development switched between two different persons ('purple' and 'green' developers).

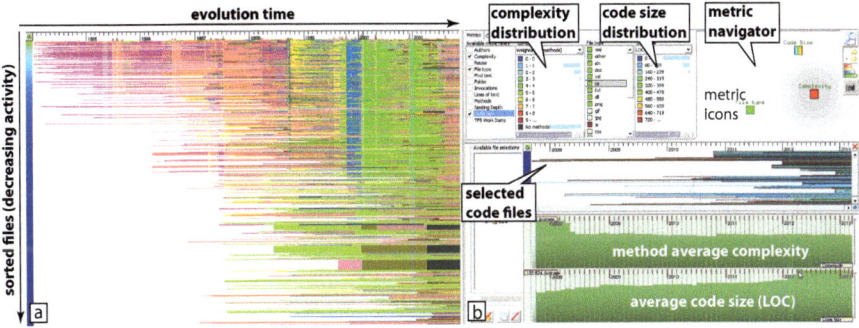

Fig. 2.10. Visualization of software activity and code quality trends in a repository

Figure 2.10b shows a detailed view from a TFS repository. Only C# source code files were selected for analysis. The top widgets show the distribution of two code metrics (complexity and size) color-coded from blue to red. These views allow easily spotting the dominant values of these metrics across a desired time or file range, which helps assessing the average code quality. The metric navigator view allows smoothly changing the color encoding used in the main file view between several metrics of interest, by dragging the red 'observer' icon between the respective metric icons, following the preset controller technique [51]. Below the file view, two graphs show the evolution in time of the selected metrics. We see, for example, that the average code size (number of lines of code per file) slightly increases, but the method average complexity first sharply decreases, then stays constant. The sharp complexity decrease is a good indicator of the presence of a refactoring event. The stability of the code quality metrics is, in turn, a good indicator that the software is well maintained.

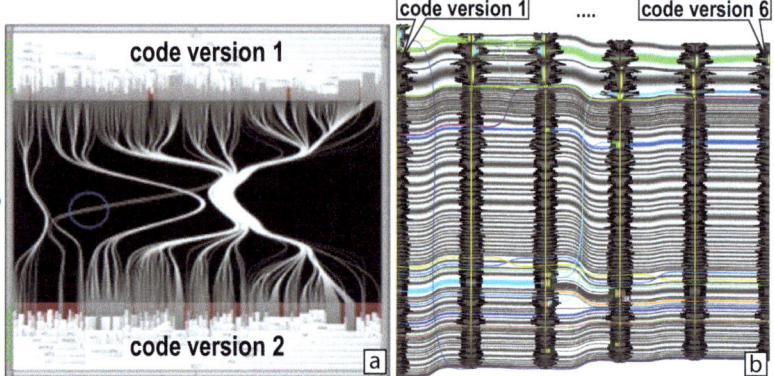

Fig. 2.11. Structural visualization of source code evolution

The aggregated views in Fig. 2.10 scale well to show the evolution of industry-size software at *coarse* file or folder levels. However, they cannot show *relations*. Figure 2.11 shows two techniques that address this challenge. In image (a), the hierarchical structures of two different versions of a software system are shown using the icicle plot technique explained earlier for trace comparison (Fig. 2.8). In contrast to traces, where x position encodes call time, the hierarchies are now permuted to place similar subtrees close to each other along the x axis [19]. Next, these subtrees are visually connected by HEB bundles. Asymmetries in the bundle structure indicate differences between the two hierarchies. The hierarchy sorting removes unnecessary bundle twists and thus increases readability. This idea is further extended by the CodeFlows tool (Fig. 2.11b). Each vertical icicle plot shows the syntactic structure of a version of a code file with the file start at top and the file end at the bottom. Similar code elements in consecutive versions, found using a syntax-aware clone detector, are connected by shaded tubes, akin to the ones used for trace comparison (Fig. 2.8). The 'flows' along the tubes indicate code refactoring events—parallel tubes show stable code, diverging ones show code insertions or deletions, and crossings show code permutations. Tube colors indicate attributes of interest, such as code element types (function, class, statement, symbol) for changed code, while gray indicates unchanged code.

Although effective to show structure change, the above techniques cannot show *association* changes. This is next achieved by extending edge-bundling to cope with dynamic graphs (Fig. 2.12). The input consists of n compound graphs mined using static analysis from n revisions in a software repository.

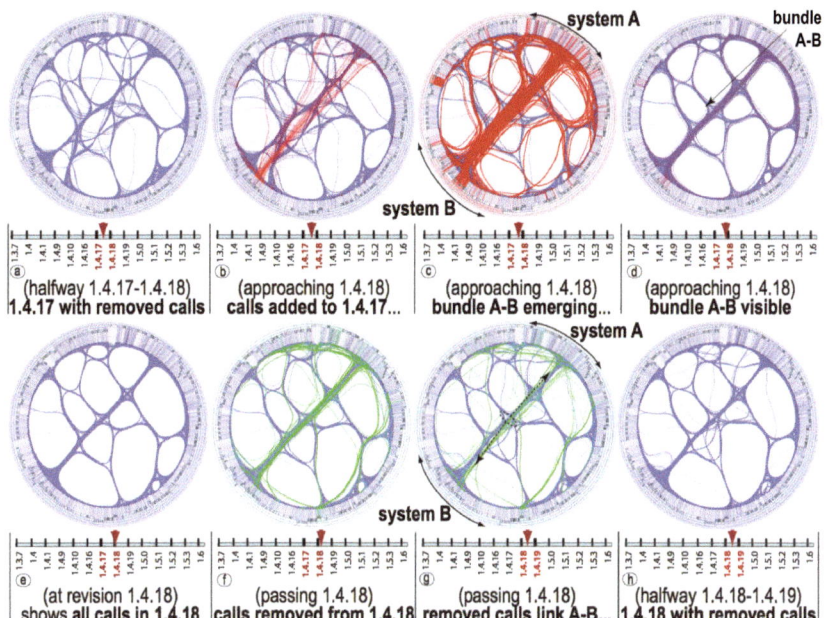

Fig. 2.12. Visualization of call graph change across multiple software versions

Here, associations are function calls. For each revision, a HEB layout is built using the associations in that revision and, as structure, the union of all entities from the n revisions. This guarantees that the radial icicle plot stays fixed for all views. Next, a continuous animation is created by smoothly interpolating corresponding edges in consecutive revisions. In parallel, appearing edges are interpolated towards their bundle, and faded in using blending, while disappearing edges are interpolated away (unbundled) and faded out. Color coding reinforces this effect: stable edges are blue; appearing edges are red, and disappearing ones are green. The images in Fig. 2.12 show eight frames from this animation between two consecutive revisions. We first see a red bundle appearing between components A and B. During the last four frames, a green bundle connecting A and B fades out. This indicates an important refactoring event between the considered revisions, when many calls connecting A and B were changed.

Visualizing changing associations in compound graphs in a single, static, image without animation is proposed by the TimeRadarTrees tool [6] (see Fig. 2.13). The top-left image shows three frames $G_1..G_3$, or snapshots, from a time-dependent compound directed graph. Hierarchy edges are orange, and associations are black. The top-right image shows the proposed visual encoding: Each node $A..E$ is represented as both a small thumbnail icon, and the correspondingly aligned sector in the large central disk. Disks are sliced in concentric rings, each one encoding a snapshot. Ring sectors in thumbnails encode the presence of an *outgoing* edge from the respective node 'revision' towards the node given by the sector's orientation. Similarly, ring sectors in the central disk encode *incoming* edges for all nodes, all revisions. Figure 2.13 (bottom) shows an application. The data is the folder-file hierarchy in the JEdit code base. Associations between two files $f_1..f_2$ indicate that f_1 was changed together (in the same revision) with f_2. Ring sectors are colored to indicate association weights, measured as number of lines co-changed between two files. Blue shows large co-changes, while gray shows no co-change. The dark-blue 'wedges' visible in the lower-right part of both the central disk and thumbnails for files $TODO.txt$ and $CHANGE.txt$ indicate that these two files co-changed over nearly the entire evolution period.

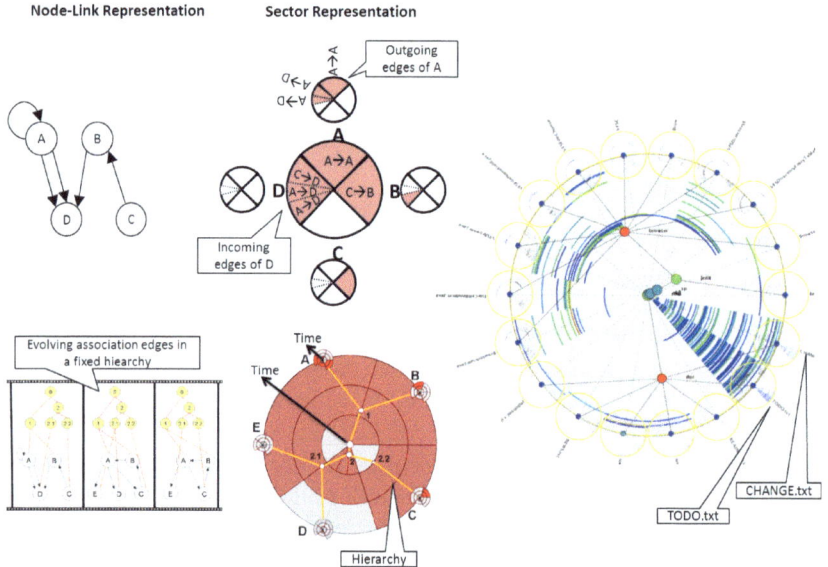

Fig. 2.13. Visualization of file co-change across multiple software versions

In contrast to the animation shown in Fig. 2.12, TimeRadarTrees unfolds time across the space (radial) dimension. As such, observing detailed evolution events is arguably easier, as all data is captured in one image. On the other hand, this method is less scalable in terms of number of nodes and relations.

2.4 Challenges and Future Directions

Summarizing our overview on multivariate graphs in software visualization, the following observations can be made.

Importance: Software structure, behavior, and evolution maps naturally to multivariate compound attributed time-dependent graphs. Such graphs have *many* attributes per node and edge, and attributes can have many different types. The *type* data is crucial to program understanding—for instance, we need to know whether a node is a function, class, or folder, and whether an edge is a call, clone, or inheritance relation, to be able to address our analysis goals.

Scalability: Being able to display *large* graphs with *many* attributes per node and/or edge is crucial to software understanding. Scalability in terms of item counts can be achieved by space-filling and dense-pixel techniques—treemaps, table lenses, timelines, SeeSoft-like views, and edge bundles. However, scalability in terms of number of attributes shown per item is quite low—most existing methods cannot show more than 2 or 3 such attributes. Linked views partially address this, but require additional user effort. Potential directions for scalability improvements are *subsampling* (drawing less items) and *dimensionality reduction* (drawing less attributes per item). However, both are challenging: For subsampling, we still do not know how to generically aggregate non-numerical attributes [11, 27]. Dimensionality reduction is a quite complex process, and can produce images which are too abstract for typical users. These issues are not unique to graphs emerging from the software engineering application domain, but important at large for any multivariate temporal graph, as discussed in more detail in Chap. 10.

Patterns: A grand open challenge in SoftVis is how to show structural, behavioral, and evolutionary *patterns*. Patterns are essential to capture (and reason about) non-trivial events in the software, such as design decisions, execution bottlenecks, and refactoring and re-architecting. However, current visualization techniques show such patterns only *implicitly*, putting the burden of detection on the user's vision. Explicitly showing such patterns would significantly guide the user towards a faster, and more profound, understanding of the studied software.

Standardization: Software visualization does not exist in a void. Many researchers have stressed that the current lack of toolchain integration (design tools, compilers, profilers, debuggers, SCM systems, and visualization tools) is a key adoption blocker of SoftVis tools in the IT industry [5, 9, 23, 33–35]. Tool communication via shared data formats [44] is helpful but not sufficient. A certain progress is visible in the last years in terms of SoftVis tools available as plug-ins to mainstream development environments such as Eclipse and Visual Studio. However, the largest majority of SoftVis tools does not follow this pattern. Separately, standardization of visual encodings used in SoftVis solutions, e.g., types of (2D *vs.* 3D) layouts, diagrams, glyphs, and color maps, is an important, but not yet covered, requirement.

In this chapter, we have presented the role of multivariate graphs in the representation and visualization of the structure, behavior, and evolution of software systems. The presented application and tool examples show that such graphs play a key role in many program understanding scenarios. Recent research in software visualization has pushed the scalability limits in terms of number of items and attributes that can be visualized. However, in the same time, we see that more challenging analysis scenarios require even more powerful tools able to display larger and more complex software patterns.

As such, developing efficient and effective techniques and tools for visualizing large, complex, multivariate, and time-dependent graphs extracted from software systems remains one of the key open challenges to software visualization. As the size and importance of the software industry grows, the creation of such tools and techniques becomes ever more necessary. In the same time, the development of such solutions for software visualization has a great potential to benefit other information visualization domains where such graphs also become pervasive.

References

1. Abello, J., van Ham, F.: Matrix zoom: A visual interface to semi-external graphs. In: Proc. InfoVis 2004, pp. 183–190. IEEE (2004)
2. Baecker, R.: Sorting out sorting (1981), 30 minute color film (developed with assistance of Dave Sherman, distributed by Morgan Kaufmann, University of Toronto)
3. Bassil, S., Keller, R.: Software visualization tools: Survey and analysis. In: Proc. IWPC, pp. 7–17 (2001)
4. Beck, F., Diehl, S.: Visual comparison of software architectures. In: Proc. ACM SOFTVIS, pp. 136–143 (2010)
5. Bessey, A., Block, K., Chelf, B., Chou, A., Fulton, B., Hallem, S., Gros, C.H., Camsky, A., McPeak, S., Engler, D.: A few billion of lones of code later: Using static analysis to find bugs in the real world. Comm. of the ACM 53(2), 66–75 (2010)
6. Burch, M., Diehl, S.: TimeRadarTrees: Visualizing dynamic compound digraphs. Comp. Graph. Forum 27(3), 823–830 (2008)
7. Burch, M., Diehl, S., Weissgerber, P.: Visual data mining in software archives. In: Proc. ACM SOFTVIS, pp. 37–46 (2005)
8. Byelas, H., Telea, A.: Visualization of areas of interest in software architecture diagrams. In: Proc. ACM SOFTVIS, pp. 105–114 (2006)
9. Charters, S., Thomas, N., Munro, M.: The end of the line for software visualisation? In: Proc. IEEE VISSOFT, pp. 27–35 (2003)
10. Corbi, T.: Program understanding: Challenge for the 1990s. IBM Systems Journal 28(2), 294–306 (1999)
11. Cornelissen, B., Zaidman, A., Holten, D., Moonen, L., van Deursen, A., van Wijk, J.J.: Execution trace analysis through massive sequence and circular bundle views. J. Sys. & Software 81(12), 2252–2268 (2008)
12. Diehl, S.: Software Visualization: Visualizing the Structure, Behaviour, and Evolution of Software. Springer, Berlin (2010)
13. Eick, S.G., Steffen, J.L., Sumner, E.E.: Seesoft—a tool for visualizing line oriented software statistics. IEEE TSE 18(11), 957–968 (1992)

14. Ferenc, R., Beszédes, A., Tarkiainen, M., Gyimóthy, T.: Columbus reverse engineering tool and schema for C++. In: Proc. ICSM, pp. 172–181 (2002)
15. Gansner, E.R., North, S.: An open graph visualization system and its applications to software engineering. Software - Practice & Experience 30, 1203–1233 (2000)
16. Goldstine, H.H., von Neumann, J.: Planning and coding of problems for an electronic computing instrument (1947), part II, volume I of a report prepared for the U.S. Army Ord. Dept., reprinted in [42]
17. Ham, F.v.: Using multilevel call matrices in large software projects. In: Proc. InfoVis., pp. 227–232 (2003)
18. Holten, D.: Hierarchical edge bundles: Visualization of adjacency relations in hierarchical data. IEEE TVCG 12(5), 741–748 (2006)
19. Holten, D., van Wijk, J.J.: Visual comparison of hierarchically organized data. Comp. Graph. Forum 27(3), 759–766 (2008)
20. Hoogendorp, H., Ersoy, O., Reniers, D., Telea, A.: Extraction and visualization of call dependencies for large C/C++ code bases: A comparative study. In: Proc. ACM VISSOFT, pp. 137–145 (2009)
21. InfoEdge: Global software industry forecast (2013), http://www.infoedge.com
22. Jones, J.A., Harrold, M.J., Stasko, J.: Visualization of test information to assist fault localization. In: Proc. ICSE, pp. 467–477 (2002)
23. Koschke, R.: Software visualization in software maintenance, reverse engineering, and re-engineering: a research survey. J. Soft. Maint. and E 15(2), 87–109 (2003)
24. Lommerse, G., Nossin, F., Voinea, L., Telea, A.: The Visual Code Navigator: An interactive toolset for source code investigation. In: Proc. IEEE InfoVis., pp. 4–12 (2005)
25. Maletic, J., Collard, M., Marcus, A.: Source code files as structured documents. In: Proc. IWPC, pp. 87–91 (2002)
26. Mens, T., Demeyer, S.: Software Evolution. Springer (2008)
27. Moreta, S., Telea, A.: Multiscale visualization of dynamic software logs. In: Proc. of EuroVis 2007, pp. 11–18 (2007)
28. Nierstrasz, O., Ducasse, S., Gîrba, T.: The story of Moose: an agile reengineering environment. In: Proc. ACM ESEC/FSE, pp. 1–10 (2005)
29. Orso, A., Jones, J., Harrold, M.J.: Visualization of program-execution data for deployed software. In: Proc. ACM SOFTVIS, pp. 67–75 (2003)
30. Pauw, W.D., Jensen, E., Mitchell, N., Sevitsky, G., Vlissides, J., Yang, J.: Visualizing the execution of Java programs. In: Proc. Inl. Sem. Revised Lectures on Software Visualization, pp. 151–162. Springer LNCS (2001)
31. Quinlan, D.: ROSE: Compiler support for object-oriented frameworks. In: Proc. CPC. pp. 81–90 (2000), see also http://www.rosecompiler.org
32. Rao, R., Card, S.K.: The table lens: Merging graphical and symbolic representations in an interactive focus+context visualization for tabular information. In: Proc. ACM Conference on Human Factors in Computing Systems (CHI), pp. 318–322. ACM Press, New York (1994)
33. Reiss, S.P.: The paradox of software visualizaton. In: Proc. IEEE VISSOFT, pp. 59–63 (2005)
34. Reniers, D., Voinea, L., Ersoy, O., Telea, A.: The Solid* toolset for software visual analytics of program structure and metrics comprehension: From research prototype to product. Science of Computer Programming 79(1), 224–240 (2014)

35. Schafer, T., Menzini, M.: Towards more flexibility in software visualization tools. In: Proc. VISSOFT, pp. 20–26 (2005)
36. Scott, A.E.: Automatic preparation of flow chart listings. Journal of the ACM 5(1), 57–66 (1958)
37. Shneiderman, B.: The eyes have it: A task by data type taxonomy for information visualizations. In: Proc. IEEE Symposium on Visual Languages, pp. 336–343 (1996)
38. Standish, T.A.: An essay on software reuse. IEEE TSE 10(5), 494–497 (1984)
39. Stasko, J., Brown, M., Price, B.: Software Visualization. MIT Press (1997)
40. Stolte, C., Tang, D., Gerth, J., Rosenblum, M., Hanrahan, P.: Rivet: a flexible environment for computer systems visualization. ACM TOG 34(1), 68–73 (2000)
41. Sugiyama, K., Misue, K.: Visualization of structural information: Automatic drawing of compound digraphs. IEEE Transactions on Systems, Man and Cybernetics 21(4), 876–892 (1991)
42. Taub, A.H.: John von Neumann: Collected Works. Pergamon Press (1965)
43. Telea, A., Ersoy, O.: Image-based edge bundles: Simplified visualization of large graphs. Computer Graphics Forum 29(3), 543–551 (2010)
44. Tichelaar, S., Ducasse, S., Demeyer, S.: FAMIX and XMI. In: Proc. WCRE, pp. 296–300 (2000)
45. Trümper, J., Döllner, J., Telea, A.: Multiscale visual comparison of execution traces. In: Proc. ICPC 2013, pp. 262–270 (2013)
46. Trümper, J., Telea, A., Döllner, J.: ViewFusion: correlating structure and activity views for execution traces. In: Proc. TPCG, pp. 45–52. Eurographics (2012)
47. USA Today: US healthcare spending (2009), www.usatoday.com/news/health
48. van Wijk, J.J., van de Wetering, H.: Cushion treemaps: Visualization of hierarchical information. In: Proc. IEEE InfoVis 1999, pp. 73–78. IEEE Press, Los Alamitos (1999)
49. Voinea, L., Telea, A.: Visual querying and analysis of large software repositories. Empirical Software Engineering 14(3), 316–340 (2009)
50. Wettel, R., Lanza, M.: Visualizing software systems as cities. In: Proc. IEEE VISSOFT 2007, pp. 92–99 (2007)
51. van Wijk, J.J., van Overveld, C.W.A.M.: Preset based interaction with high dimensional parameter spaces. In: Post, F., Nielsen, G., Bonneau, G. (eds.) Data visualization – State of the art, pp. 391–406. Kluwer (2003)

3
Multivariate Social Network Visual Analytics

Chris Muelder, Liang Gou, Kwan-Liu Ma, and Michelle X. Zhou

One of the key research topics in Social Science and Sociology is to understand and analyze various social networks. Like any other types of networks, a social network consists of a set of nodes and links. Here, each node often represents a social entity, such as an individual or a group, and each link represents a particular relationship between two social entities. In a *multivariate* social network, each node/link can be associated with a set of properties, or there can even be multiple sets of heterogenous nodes or edges.

3.1 Data Characteristics

In addition to understanding the behavior of individual social entities, Sociology is also concerned with the behavior of *groups*, in particular, how these groups interact with each other [49]. In this context, a multivariate social network is composed of the entities of groups. The connections between them depend on the task being pursued and the information that is available, but are generally a set (or multiple sets) of relationships between the entities. These relationships can be directed or undirected, weighted or unweighted. Additionally, the nodes can carry any additional properties.

In particular, social networks are imbued with a number of properties. The size and complexity of the topology itself can be overwhelming for many traditional approaches. Additionally, both the nodes and links can carry any number of properties, including nominal, ordinal, and continuous measures: for instance, nodes can often be broken down into classes both ordered (age, grade, etc.) and unordered (gender, race, etc.), or there can be multiple classes of edges on the same set of nodes (e.g., both friendship and aggression ties between the same group of actors), or the nodes or edges could contain weights, or even multiple weight metrics, and the edges could be directional if the network is not symmetric. Also, many social networks evolve over time, so while static analysis can reveal some insights, in many cases the evolution of dynamic social networks could be of importance.

To make things more complicated, a multivariate social network may not be homogeneous as it may contain multiple types of nodes, which represent both individuals and groups. For example, an enterprise social network may include nodes representing individual employees as well as those representing companies or organizations that are customers, suppliers, or partners of the enterprise.

All of the mentioned properties are often compounded by a difficulty in acquiring the network data. Next, we briefly describe different approaches to acquire social network data.

3.1.1 Traditional Data Collection

Traditionally, social network data are acquired by polling small groups of people, where people report on their social ties via questionnaires. This introduces numerous points for the introduction of uncertainty. Not only do the subjects' responses depend on the questions asked, but even on how they are worded. The accuracy of the data also relies on the honesty of the subjects. In addition, temporal resolution of such networks is extremely low, since it is very difficult to get a large number of subjects to willingly and dutifully fill out one questionnaire, let alone repeated (e.g., weekly, monthly, or even annual) questionnaires. To address these challenges, alternative data acquisition approaches have been considered, particularly in social network analysis of non-human actors such as herds of animals. In these cases, the actors are tagged and tracked, but such tracking obfuscates most details of the network, as the animals are unable to communicate the specifics of their relationships, e.g., a proximity test could determine two animals met, but not inherently determine if the meeting was amicable or hostile. In either case, most traditional social network data collection methods result in data sets that are small, incomplete, noisy, vague, or even unreliable.

3.1.2 Data Collection from Online Social Media Networks

Conversely, the advent of online social media, such as Facebook and Twitter, has created the ultimate data source for social network analysis, as an incredible number of users are readily willing to divulge both explicit and implicit social connectivity information in order to benefit from the service that social media provides. This has resulted in an explosion of social network data in recent years. The result of this is that sociologists now often have to deal with massive data challenges, such as handling extremely large networks or real-time trend detection and analysis. But the emergence of social media has also introduced privacy issues that often limit the third party access to these networks and also limit what can be done with them.

In parallel to collecting social network data from public social media sites, an alternative approach is to build social networks from people's communication data, including emails, online chats, phone calls, and meeting invites, especially within the context of enterprise [59]. Not only can such data connect one social entity to another, but they can also be used to characterize the relationships between any two connected entities, including their tie strength, topic of interest, and style or type of communications. Furthermore, such information can the be used to better understand the characteristics of an individual's as well as an organization's social network [59].

3.2 Task Characteristics

Given a multivariate social network, the typical tasks of understanding such a network are to analyze its different social entities, and the properties of the entities or the network as a whole.

3.2.1 Understanding Social Network Nodes

As mentioned in the previous section, a node of a multivariate social network represents an individual or a group, which is often associated with a set of traits describing the individual or group. With the emergence of social media and advances in data analytics, much information can be inferred from one's social media footprints to describe various traits of the individual or group. In particular, there is much research on understanding various traits of an individual, from demographics to political orientation to personality traits [21, 39, 47]. Similarly, there is also much work on extracting the properties of a group, including aggregated properties of a group such as the level of expertise [46] or the discovery of latent communities/groups along with their properties [48, 58].

3.2.2 Understanding Social Network Links

Since a link represents the relationship between two social entities, understanding a link is often to characterize such a relationship (e.g., type and strength) and predict its properties (e.g., likelihood to last). The relationship between two social entities can be characterized in many different ways. For example, between two individuals, such a relationship can be used to describe what, how, and when such a relationship is established [59]. Besides understanding the characteristics of a relationship, there is also research on predicting the properties in particular the existence of a particular relationship between two entites [26].

3.2.3 Understanding Social Networks

Understanding a network as a whole is a complex task as it depends on the purposes of the analysis as well as the analytic technologies. Because of the challenges, visualization is often developed to accompany the analytics technologies to help users better understand various properties of a network.

Graph-Based Analysis

Many sociological research works draw on graph analytic algorithms and statistics. These analyses can range in scale from looking at small scale patterns such as dyads (pairs of entities) [49] or triads (groups of 3) [17], to centrality metrics that measure nodes' importance to the network as a whole [18], and up to large scale analysis of the high-level relationships that find and compare large groups of entities and how they interact.

Sociograms Analysis

One key element of social network research has been visualization of node-link diagrams, which sociologists often refer to as "sociograms" [19]. While statistical metrics can be quite succinct, it can be difficult to know a priori what metric will produce the right result, and it can be difficult to directly verify that the results are correct. Pictorial representations of social networks can help to both directly communicate the content of the network such as structural patterns, as well as to guide and confirm the choices of statistical metrics. Nevertheless, traditional visual diagrams of social networks often suffer from a range of problems, the most common of which being the high density of edges and complex structures in large networks, yielding sociograms that often appear as indecipherable clouds of nodes and edges.

3.3 Examples of Technologies

3.3.1 Clustering

Another way to simplify large, complex networks is to cluster tightly connected groups of nodes together and consider the resulting abstracted supernetwork. Many current clustering algorithms are based on the modularity metric, such as the Louvain clustering method [3] or the "Fast Community" clustering algorithm of Clauset, Newman, and Moore [8]. These clustering algorithms have been shown to be effective on real-world networks, as the modularity metric is demonstrably comparable to force directed energy functions [45]. Modularity is a metric that evaluates a specific proposed clustering of a network by measuring the density of cluster interiors and the sparsity of inter-cluster connections. Specifically, given a network with a proposed clustering, the modularity Q is defined as:

$$Q = \frac{1}{2|E|} \sum_{i,j} \left[A_{i,j} - \frac{k_i k_j}{2|E|} \right] \delta_{i,j} \qquad (3.1)$$

where $|E|$ is the number of edges in the network, k_i, k_j are the degrees of nodes i and j, $A_{i,j}$ is 1 if there is an edge between nodes i and j and 0 otherwise, and $\delta_{i,j}$ is 1 if nodes i and j are in the same cluster and 0 otherwise. Recent efforts have also been shown to make such approaches produce more balanced hierarchies [31] or to parallelize the clustering calculation [36].

3.3.2 Network Centralities

Centrality metrics are commonly applied to the analysis of social networks, such as Eigenvector [6, 35], Markov [57], Betweenness [18, 38], and Closeness [32, 44] centrality. Each of these measure vertices' overall importance with respect to the whole network. Rather than basing the importance of a node solely on how many connections it has, eigenvector centrality also takes into account the weights of connections to other nodes; a single connection to a highly important node can carry more weight than many connections to nodes of low importance. Eigenvector centrality sensitivity extends this notion to derive the importance of nodes relative to each other.

3.3.3 Centrality Derivatives

While centrality gives one value per node, centrality sensitivity analysis measures a vertex's importance to the structure of the network relative to other vertices in the graph [9]. These metrics are essentially derivatives of centrality, and as such can be calculated similarly for any type of centrality. To calculate a reference node's sensitivity to a target node, the reference node's initial centrality is calculated, each edge of the target node is removed one at a time, and the centrality of the reference node is recalculated after each removal. The negative changes in centrality of the reference node give a measure of how important the target node is to the reference node—in other words, how sensitive the reference node's centrality is to the target node. For instance, if removing a target node's edges results in large decreases in the reference node's centrality, then the reference node is said to be highly sensitive to the target node—that is, the target node has high importance relative to the reference node. This can be summarized in the following equation [9]:

$$\frac{\partial x}{\partial t_i} = -Q^+ \frac{\partial Q}{\partial t_i} x \tag{3.2}$$

where x is the centrality, t_i is the degree of vertex i, Q is the subtraction of the identity matrix from the adjacency matrix of the network ($Q = A - I$) (A is the adjacency matrix, and I is the identity matrix), and Q^+ is the pseudoinverse of Q.

One application of these sensitivities is to evaluate the roles of edges in the graph. If two nodes impact each other negatively, then they have a competitive relationship, whereas other nodes have mutually beneficial relationships. This can be shown as simply as using color, as in Fig. 3.1. Alternately, since every node has a centrality derivative with respect to every other node, centrality sensitivity can be thought of as a complete, weighted network. From

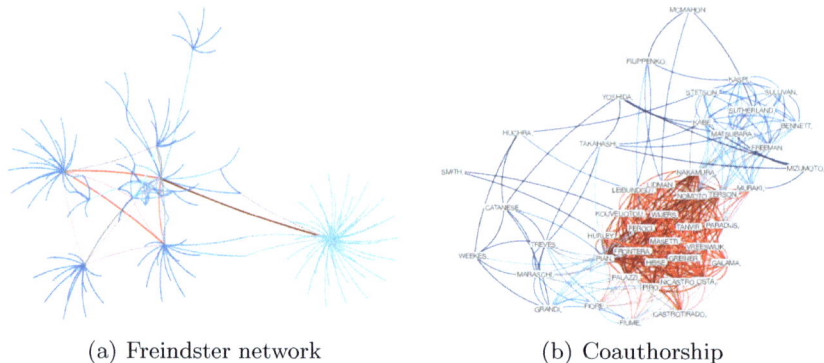

(a) Freindster network   (b) Coauthorship

Fig. 3.1. Centrality sensitivity analysis can indicate how collaborative (blue) or competitive (red) relationships are along edges in the network

this network, it is possible to derive a skeleton network based on edge existence, high centrality derivatives, and overall connectivity (e.g., using a spanning tree). This skeleton network can then be thresholded to be as sparse or as dense as needed, and can be used for a wide variety of purposes, such as simplifying/clarifying layouts (as in Fig. 3.2), visualizing only the most important connections, or finding important relationships between nodes with no direct connections.

3.3.4 Traditional Network Layouts

One key task in creating visual images of networks is to determine the appropriate geometrical layout of the nodes and edges. There are several well-defined criteria for assessing the accuracy and validity of a particular graph layout [13]. Some common criteria [2, 4] include, but are not limited to:

1. edges of the same approximate length;
2. vertices distributed over the area;
3. reduction of the number of edge crossings.

Nevertheless, optimization of such criteria can be intractable and often contradictory [4]. For surveys of many modern graph layout algorithms see Battista, Eades, Tamassia, and Tollis [55] or Hachul and Jünger [24].

The most traditional and commonly used layout algorithm for social network analysis are force-directed layouts [33], often referred to as "spring embedders" [15]. In this well-known procedure, nodes in a network graph are positioned iteratively, where the edges connecting them are treated like springs that push and pull on them until the system converges to an equilibrium. However, spring embedder techniques do not always scale nicely to large graphs [5]. Thus, a common problem that faces many existing visualizations of large social networks (most of which use force-directed layouts) is that they often result in a tangled mess of incomprehensible lines; this is often referred to as the "hair-ball" problem (Fig. 3.2(a) shows an example).

Other approaches have been developed with the goal to improve network layout in terms of quality and algorithmic efficiency, especially for large graphs. One such technique [4] is based on a variant of dimension-reduction methods, referred to as multidimensional scaling [10], in which the goal is to minimize stress. In this approach, the purpose of stress minimization is to determine positions for every node such that the Euclidean distances in the n-dimensional space resemble the given distances between the nodes, as determined by graph-theoretic measures, such as the shortest paths (i.e., geodesics). However, such geodesic based layouts tend to fail on networks with small diameter, as is common among social networks.

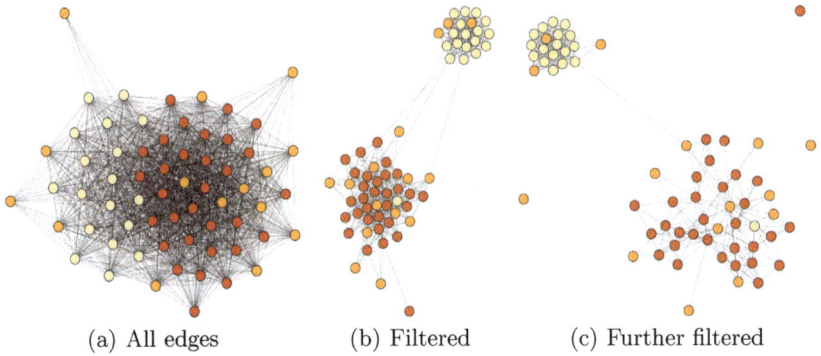

(a) All edges (b) Filtered (c) Further filtered

Fig. 3.2. MIT reality data set. Trying to lay out the whole network can yield an unintelligible hairball (a), but filtering out the less important edges via sensitivity analysis reveals two clusters (b), and further filtering starts to dissolve one of them while the other remains strong (c).

3.3.5 Improved Network Layouts

One method for improving the layout of dense social networks is to trim the network of its less essential connections to reduce it to a core network consisting of just the most important connections. A naïve way to do this would be to simply take a spanning tree of the network itself, but this is not always ideal for preserving the centralities of the nodes, which sociologists are often concerned with. Instead, the edge filtering can be weighted according to the centrality derivatives, so that the edges that are removed are the ones that affect the centralities the least. This produces a core network that preserves as much of the critical structure of the network, which can then be used to create an improved layout of the graph that reveals more detailed structures, as in Figs. 3.2(b) and 3.2(c). Once this reduced network is laid out, the original edges can optionally be reintroduced.

As social networks tend to exhibit strong community structures, cluster-based layouts based on hierarchical structures have proven useful, such as the treemap layout [41] or space-filling curve based layouts [42], as shown in Fig. 3.3. A treemap defines a hierarchical decomposition of screen space, where the whole screen is recursively subdivided according to the tree, i.e., the root of the tree takes up the whole screen, each branch subdivides the screen at each level of the tree, and finally each leaf of the tree is allotted its own region of the screen. When applied to a graph's clustering hierarchy, each node in the graph is a leaf in the hierarchy, and can thus be placed in the corresponding region to define the layout. In the space-filling curve layout, the nodes are ordered in 1 dimension, and then mapped to the screen using a recursively defined fractal curve, such as the well known Hilbert or Gosper curves. Any such clustering-based layout can provide clear boundaries between communities—particularly when combined with edge bundling techniques. And since a clustering is already computed, hierarchical edge bundling is a good fit [28].

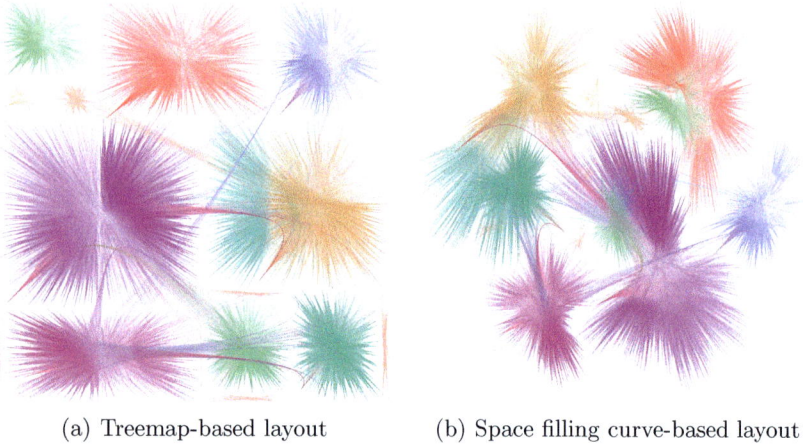

(a) Treemap-based layout   (b) Space filling curve-based layout

Fig. 3.3. Clustering-based graph layouts using trees (a) or space filling curves (b) can be used to show explicit separation between communities

3.3.6 Multivariate Social Networks

Sometimes, merely improving the layout algorithm is insufficient for showing particular aspects of a network. Specifically, social networks can often be divided into groups according to discrete properties besides connectivity, such as gender, race, school grade, or others. However, the density of ties in most traditional node-link diagrams make it difficult to distinguish in inter-group patterns from intra-group patterns, as in Fig. 3.4(a). One approach to address this is a modified radial representation that arranges nodes according to

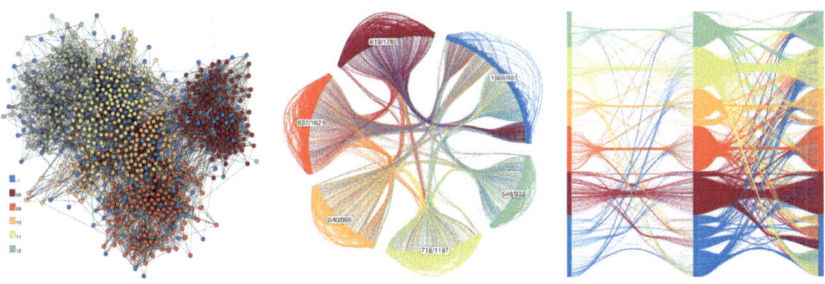

(a) Traditional layout (b) Radial layout, divided by node category (grade) (c) Parallel layout, with multiple edge sets (aggression vs. friendship)

Fig. 3.4. Some networks can be divided up categorically, with multiple node categories or edge sets. Here we show a social network of students colored by grade. Traditional node-link diagrams (a) can be too cluttered to read, but explicitly dividing nodes by category (b) or showing multiple edge sets in parallel (c) can better show how these categories or edge sets interact. (Images from [12] with permission.)

categorical properties in addition to connectivity, as shown in Fig. 3.4(b). Nodes are placed around a circle, grouped into discrete arcs based on the selected data attribute, and ordered within each group by connectivity with the use of modularity clustering. This new representation also delegates the two kinds of connections to separate regions of space: intra-group edges are displayed outside the circle while inter-group edges are drawn in the middle. The label on each group shows the number of inter-group and intra-group connections, respectively.

In addition to node categories, social networks can also contain more than one kind of edge, defining two or more unique networks on the same set of nodes. In such cases, a layout that is good for one set of edges might not be good for another. Alternately, with one unified layout, sparser networks may get lost inside denser ones. One approach to address this is a representation based on n-partite network layouts, where layers of nodes are laid out parallel to each other, similar to the dynamic graph approach of Burch et al. [7]. This concept has been applied to multiple edge sets on the same set of nodes by replicating the nodes in each layer, and considering each edge set as a bipartite graph from the full set of nodes to a duplicate set of nodes, which creates an n-partite network where n is one more than the number of edge sets. This n-partite network is then laid out in a series of columns by evenly spacing the nodes in each column. Edge directionality is also shown in this representation, since all edges proceed from left to right. While hierarchical layouts such as Sugiyama [53] or Dig-Cola [14] could be used, here each layer of nodes is identical, and thus it is more natural for each column to have the same ordering. Thus, the afore calculated categorical modularity clustering is applied to cluster the nodes, and the resulting clustering is traversed to define a universal ordering. An example of this parallel layout is shown in Fig. 3.4(c).

3.3.7 Dynamic Social Networks

In real world applications, social networks are often intrinsically time-varying: New friendships can be made, or old friendships lost. While the problem of visualizing static networks has been studied quite extensively, work on dynamic network visualization is less mature.

A common method for visualizing dynamic graphs is to animate the transitions between time steps. This approach yields dynamic visualization with nodes appearing, disappearing and moving to produce a readable layout for each time step. Alternatively, multiple time steps can be statically placed next to each other using "Small Multiples" [56]. This eases the comparison of distant time steps but limits the area devoted for each time step which reduces the legibility of each graph. Archambault et al. [1] have done an empirical study to compare the advantages and drawbacks of these approaches (i.e., "Animation" vs. "Small Multiples"). In either case, when creating a node-link diagram for a dynamic graph, not only does the layout need to consider graph topology, but also the stability between time steps. Hu et al. [30] proposed a method based on a geographical metaphor to visualize a summary of clustered dynamic graphs. An alternate visualization approach for dealing with dynamic large directed graphs is to directly represent time as an axis. In the work of Burch et al. [7], vertices are ordered and positioned on several vertical parallel lines, and directed edges connect these vertices from left to right. Each time-step's graph is thus displayed between two consecutive vertical axes.

Storyline visualizations have become popular in recent years for showing evolution of interactions such as clusterings or networks [54]. Sallaberry et al. [51] use a globally optimized dynamic graph clustering approach to both extend the SFC layout method [42] and create a storyline-like timeline representation of the network. An example of such a timeline is shown in Fig. 3.5.

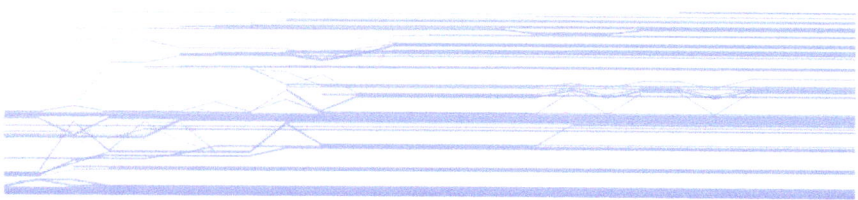

Fig. 3.5. Evolution of a small social network collected off the Rimzu social media site

3.3.8 Egocentric Approaches

Due to screen and retinal resolution limits, and a psychological limit on attentiveness, there is a finite maximum amount of information that can be conveyed by any one view. Thus, as datasets get even bigger, an overview of the dataset will show proportionally less and less of the underlying data. As a means to address this, researchers have introduced several bottom-up techniques, which bypass or supplement the overview with a detailed view that starts at the lowest level of the data (i.e., a single selected node and its immediate context). Additional relevant nodes and connections are revealed only on demand, based on graph structure or specialized degree-of-interest (DOI) functions.

"Link Sliding" and "Bring & Go" are two such DOI functions for navigating large networks [40]. Heer and Boyd [27] presented a visualization method which only shows a focus node's neighboring nodes up to a certain level. Similarly, Elmqvist and Fekete [16] described a bottom-up system based on hierarchy traversal methods. These methods are useful when the inherent graph structure is more important than other properties for the task at hand. For other applications, where node/edge attributes are the focus of analysis, researchers create specialized DOI functions. Furnas [20] introduced a DOI function to evaluate the importance of a selected node based on distance and a priori interest. Van Ham and Perer [25] extended this function to operate on embedded attributes and graph topology, as well as user-generated search actions. Crnovrsanin et al. [11] combine this concept with an interaction history based importance similar to Amazon's item-to-item collaborative filtering [37]. The result of this is a visual recommendation system that takes into account not only the underlying topology, but also the users' interaction histories. An example of a path in a user's exploration is shown in Fig. 3.6(a).

In dynamic networks, not only will importance depend on the interactive selection of focal points, but also on the temporal history of the network. Muelder et al. [43] have extended the DOI functions for dynamic networks by using computing a DOI that takes into account not only static topology, but also temporal topological history and interaction history. This is then used along with dynamic clustering to create focused, egocentric storylines, as shown in Fig. 3.6(b).

3.3.9 TreeNetViz: Revealing Patterns of Networks with Hierarchical Attributes

This sample technology demonstrates a new visualization technique, TreeNetViz [22], to help users understand a network with hierarchical attribute information. This technology is built up on a TreeNet graph, a type of multivariate network in which node attribute has hierarchical structure. For example, as shown in Fig. 3.7, a subgraph of a scientific co-author network in Fig. 3.7a has node attribute of affiliation, such as country, university and department,

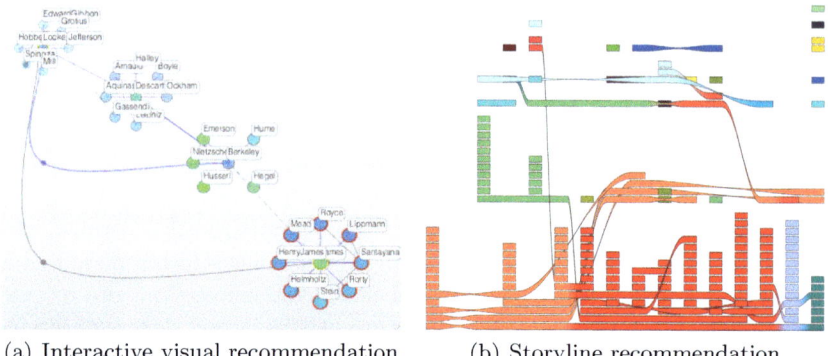

(a) Interactive visual recommendation (b) Storyline recommendation

Fig. 3.6. Using a recommendation system to focus on the neighborhood around a single focal node over time can show dynamic context of changes in that individual's relation to the network. In static networks, this change can be from interactive focal point changes as the user explores (a). In dynamic networks, this change can also be due to the evolution of the network itself over time (b).

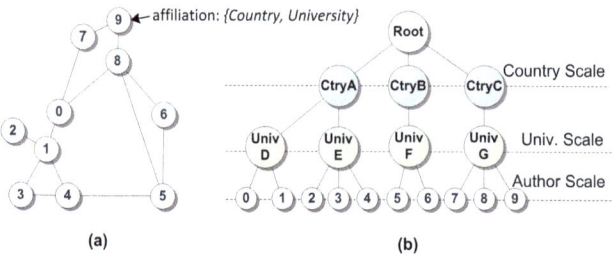

Fig. 3.7. An example of a TreeNet graph. It includes (a) a scientific co-author network, and (b) node affiliation attribute with a hierarchical structure.

and the affiliation attribute has a hierarchical structure shown in Fig. 3.7b. This type of graph, a special type of multivariate network, is called TreeNet graph.

Analysis of this type of network is not a trivial task. It is important to analyze the connectivity, centrality and path patterns at different levels aggregation on the node attribute. For instance, to fully understand the scientific co-author network shown in Fig. 3.7, the collaboration activities can be analyzed through different entities, from individual authors to multiple universities to international collaborations. Th analysis is achieved by aggregating network connections at different levels of node attribute hierarchy. This type of analysis enables us to understand an individual's social activities at different affiliation levels [34].

TreeNetViz Design. TreeNetViz is designed to support various multivariate networks analysis at different levels of node hierarchy for a TreeNet graph.

TreeNetViz uses a Radial, Space-Filling (RSF) [52] technique to show a tree structure of the node attribute in the TreeNet graph (Fig. 3.8a). It then uses a circular layout for an aggregated network and places the aggregated network over the RSF tree (Fig. 3.8b and c). To reduce visual cluttering, it adopts an edge bundling technique based on [29](Fig. 3.8d). It also includes an algorithm to improve circular node placement to reduce the edge crossings with the consideration of various constraints.

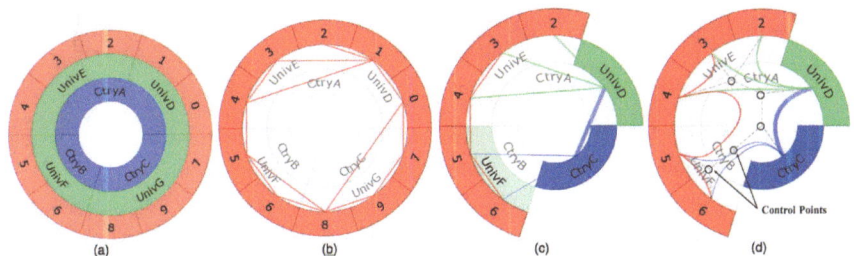

Fig. 3.8. TreeNetViz Visualization Design: (a) a Radial, Space-Filling (RSF) layout of the node attribute structure; (b) the optimized circular layout of the network overlaid on RSF tree; (c) a RSF circular layout of an aggregated network; (d) the view after edge bundling

Treenetviz also includes rich interactions to support network analysis tasks at different levels of aggregation. It enables users to observe network patterns (connectivity, centrality, and reach) among entities of the same type (e.g. the collaboration patterns among all universities or countries in previous example) by controlling the view level. It supports arbitrary aggregation of network by expanding and folding node sector in the visualization. It also enables users find the short paths among nodes of interest in aggregated networks.

An example application of TreeNetViz. A TreeNetViz example is presented to help people understand collaboration patterns among researchers in a co-author network. The collaboration network was extracted from MedLINE research articles published from 2006 to 2010 in the area of diabetes at University of Michigan. The data set includes 614 articles, 847 authors and 2,498 co-author relationships. 10 college-level nodes and 90 department-level nodes are identified.

Fig. 3.9 shows the visualization results of collaboration patterns at three different levels of colleges (Fig. 3.9a), departments (Fig. 3.9b), and individuals (Fig. 3.9c). With this visualization, people can understand network patterns at different scales from the perspectives of the power and status of collaboration resources, and the access control to social groups and individual authors.

TreeNetViz also presents patterns how social actors collaborate with each other from different scales. As shown in Fig. 3.10a, the collaboration patterns of researchers in the "Biochemistry Dept" with other departments in

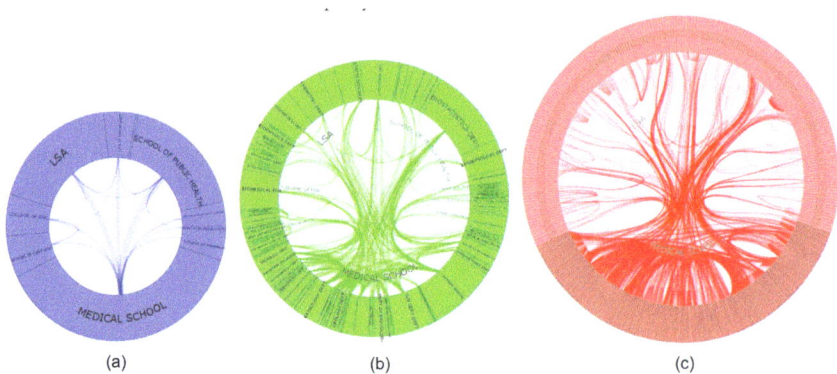

Fig. 3.9. TreeNetViz visualizes collaboration patterns at three different levels: collaborations among colleges(a), departments(b), and individuals(c)

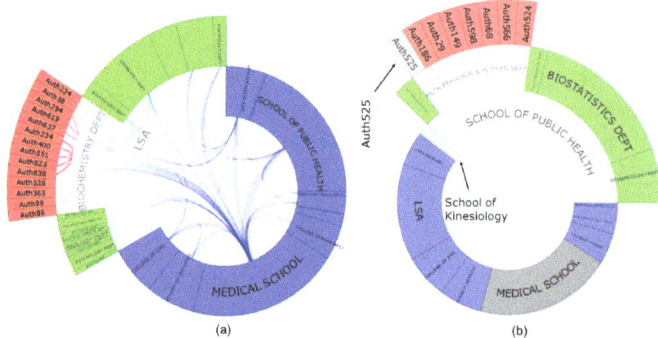

Fig. 3.10. (a) Collaboration patterns across different levels of entities of colleges, departments and individuals; (b) A critical path connecting two colleges by the researcher "Auth525"

"LSA", and other colleges are presented. It is also helpful to identify important people connecting different organizations. Figure 3.10b shows the researcher "Auth525" connects two organizations: "School of Kinesiology" and "Medical School".

3.3.10 SocialNetSense: Making Sense of Multivariate Social Networks

While TreeNetViz is a specific visualization technique to represent and help people to explore a specific type of multivariate network, SocialNetSense [23], on the other hand, is a visual analytics tool to support different analysis tasks on social networks with rich node attributes. SocialNetSense integrates different visualizations of multivariate social networks and supports the analysis process with a sensemaking approach. Here, the social network with rich node attribute information, such as TreeNet graph, is a type of multivariate network of interest.

Sensemaking approach for visual analytics. SocialNetSense adopts a sensemaking approach to support the visual analytics of multivariate social network. Sensemaking is a process to iteratively construct and refine a representation or understanding of data and fit data with the representation to meet the requirements of a task [50]. There are several important sensemaking tasks on social networks with rich node attribute information including understanding network features, the social attribute features and the hybrid features of network and attribute.

Figure 3.11 shows the sensemaking framework for multivariate social network visual analytics. The framework consists of a network exploring loop and a representation building loop. In the network exploring loop, users can explore social attribute features, network features, and hybrid features to collect information based on their tasks and existing knowledge. Various metrics (such as degree, betweenness and closeness), plots (such as degree distribution) and visualization tools are implemented to help users to explore these features. On the other hand, in the representation building loop, users process and comprehend the information collected, build and revise their representation of the data.

The two loops interact with each other with bottom-up and top-down processes. In the top-down process, representations are used to guide users' exploration to look for new evidences. In the bottom-up process, information of interest are collected as evidences to confirm or dis-confirm the representation. As more evidence are collected, representations can be revised and even re-constructed.

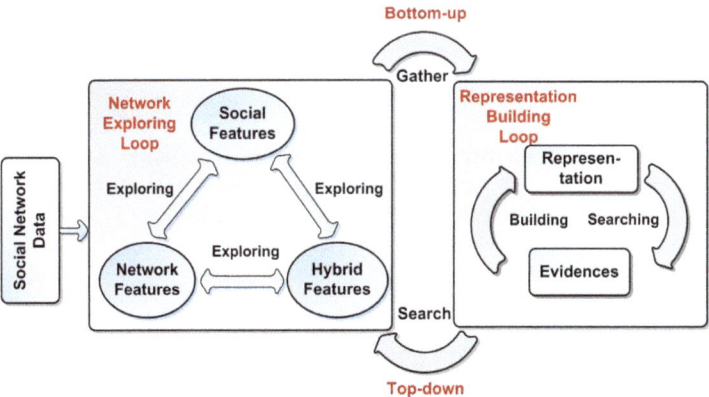

Fig. 3.11. A sensemaking framework for multivariate social network visual analytics in SocialNetSense

SocialNetSense User Interface. Guided by the sensemaking framework, the SocialNetSense user interface includes three main components: a Network Exploring Space (NES), a Representation Building Space (RBS), and a process view.

The interface of NES is shown in Fig. 3.12, including two main panels: a visualization view (Panel 1) showing social networks along with hierarchical social structures, and a control panel (Panel 2) offering a set of analytical tools, such as different aggregation metrics of connectivity and centrality, analytic plots and searching function. View manipulation tools, such as zooming, panning and layouting, are provided in the tool bar above Panel 1. It uses multiple visual representations of networks to facilitate the exploration of social, network and hybrid features. Node-link diagrams (Panel 1 in Fig. 3.12) are used to show social features and network features, and TreeNetViz is used to show hybrid features of aggregated networks over node attributes.

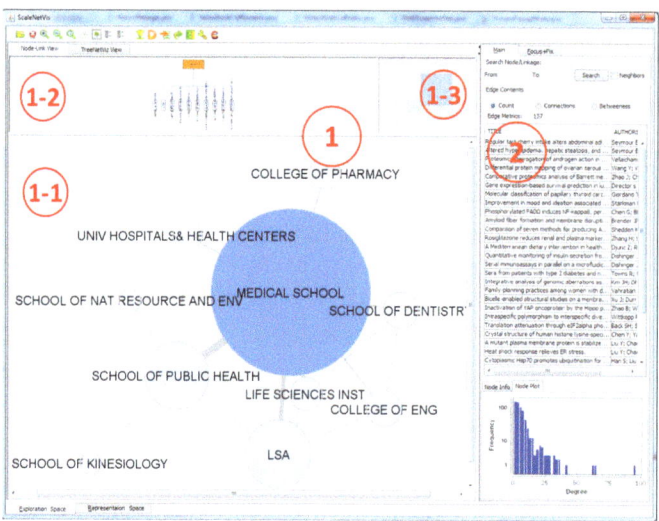

Fig. 3.12. Network Exploring Space (NES) in SocialNetSense: (a) Coordinated node-link views, including a network visualization (Panel 1-1), a tree visualization (Panel 1-2) for social hierarchy, a network overview panel (Panel 1-3), and a control panel (Panel 2) with analytical tools

In RBS, users can organize the evidence collected from the NES to create their representations using editing functions. Figure 3.13 shows the user interface of RBS. The interface consists of a editing space (Panel 1), a process view (Panel 2) and an element list view (Panel 3). The editing space enables users to collect visualization elements from the node-link view and the TreeNetViz view, network metrics of size, centrality, betweenness and closeness, and also plots from the NES. It also provides functions such as grouping/ungrouping, note-taking, and element-linking to build representation.

3.3 Examples of Technologies   53

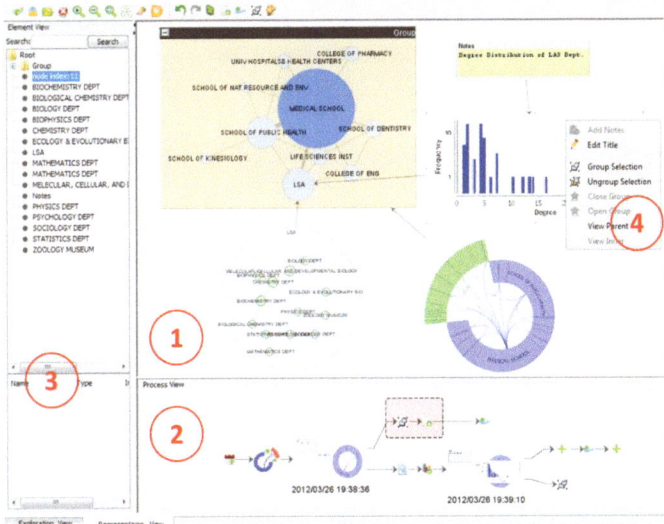

Fig. 3.13. Representation Building Space (RBS): Panel 1 is the main working space; Panel 2 is the process view; Panel 3 lists the elements in the working space; Panel 4 shows tools of representation building, such as add note, group/ungroup elements and link element

Sample analysis with SocialNetSense. The same data set of a diabetes researcher collaboration network is used to demonstrate how SocialNetSense supports the visual analytics of multivariate social networks.

With SocialNetSense, users can build their understanding of the collaboration patterns, such as the power and status of social actors and collaborations, at three different levels (colleges, departments, and individuals). In Fig. 3.14, an example representation is shown for the main network patterns at the level of colleges composed by a user. The strong collaboration among "Medical School", "LSA" (Literature, Science and the Arts) and "Public Health" is captured with detailed notation of network metrics, plots, and co-authored articles. Similarly, it can also help users understand cross-scale patterns such as collaboration among the departments in "LSA" with other colleges.

With SocialNetSense, users can have comprehensive understanding of the analytics process. Figure 3.15 shows how a user makes sense of the network to identify an important actor (Author 525) acting as a "boundary spanner" to connect "Medical School (MS)" and "School of Public Health (SPH)". Compared with the visualization result shown in Fig. 3.10b, SocialNetSense shows the intermediate results and reasoning process.

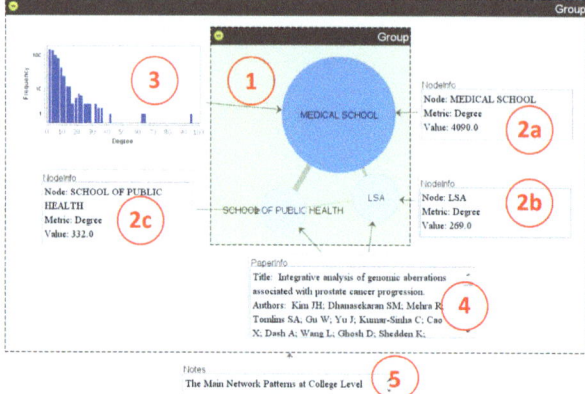

Fig. 3.14. A sample representation of the main network patterns at the level of college

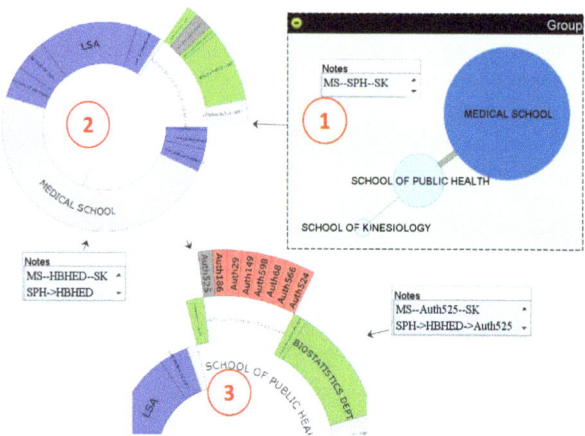

Fig. 3.15. A representation of a boundary spanner connecting "Medical School (MS)" and "School of Public Health (SPH)"

3.3.11 Summary

Many works have proposed methods to accomplish these tasks. Dyadic and triadic analyses often rely on basic statistics, but other metrics require more complex algorithms. There are a number of centrality metrics, each with their own strengths and weaknesses, and a number of additional metrics derived from centrality metrics. Community-scale analyses depend on clusters, so there are also a substantial number of clustering algorithms.

As node-link diagrams are traditional for visually inspecting social networks, node-link diagram layout algorithms are intrinsically applicable. But social networks exhibit certain properties that make many layout algorithms less useful. So, there are additional approaches to improve layout algorithms for social networks, based on relevant statistical analyses such as centralities or clusterings. Some visual approaches even incorporate semantic information, such as node categories [22] or multiple edge sets. There has also been recent work in extending visual analyses to dynamic social networks. And finally, as the size of social networks available to researchers has grown incredibly in recent years, bottom-up visual analytic approaches, such as recommendation and sensemaking-based systems [23], are becoming increasingly popular.

3.4 Challenges and Future Directions

While much work has been done on the visualization and analysis of social networks, many of the key challenges are only getting more important. The size of social network data available has exploded in recent years due to social media, and continues to grow every year. Many of these networks generate complex data in real-time, and real-time analysis offers many unique opportunities and challenges. The kind of information in social networks can also be quite varied, such as social network messages from different devices, different locations, and different social media sites, and may contain various meta-data that could potentially improve analytic results. The validity of the information must also be considered, as most social networks rely on the honesty of the users, and are potentially vulnerable to wildly inaccurate input, missing data, or even spam. And lastly, there is much that can be done to improve the analytic insights to be gained from previous data that has already been collected.

As such, there are numerous opportunities for future work in this area. As there are many social media sites, finding new ways of combining and analyzing networks from various sources would be beneficial to creating a more complete picture of the underlying social trends. Producing useful analytic results while preserving the privacy of the subjects is also important, as many users would be more willing to provide accurate information if they trust the privacy policies. But the resulting networks will still have uncertainty, whether due to the data being sanitized to protect privacy or the users omitting data to protect their own privacy. So incorporating further uncertainty metrics to either measure the validity of the input data or conjecture missing information would aid in improving the analytic results. And lastly, even when a visual analytic process produces an insight, it is important for the analyst to convey the underlying derivation, as tracking and analyzing the provenance of insights is critical for improving the analytic process.

References

1. Archambault, D., Purchase, H.C., Pinaud, B.: Animation, small multiples, and the effect of mental map preservation in dynamic graphs. IEEE Transactions on Visualization and Computer Graphics 17(4), 539–552 (2011)
2. Bertin, J.: Semiology of graphics. University of Wisconsin Press (1983)
3. Blondel, V., Guillaume, J., Lambiotte, R., Lefebvre, E.: Fast unfolding of communities in large networks. Journal of Statistical Mechanics: Theory and Experiment 2008(10), P10008 (2008)
4. Brandes, U., Indlekofer, N., Mader, M.: Visualization methods for longitudinal social networks and stochastic actor-oriented modeling. Social Networks 34(3), 291–308 (2011)
5. Brandes, U., Pich, C.: An experimental study on distance-based graph drawing. In: Tollis, I.G., Patrignani, M. (eds.) GD 2008. LNCS, vol. 5417, pp. 218–229. Springer, Heidelberg (2009)
6. Brin, S., Page, L.: The anatomy of a large-scale hypertextual Web search engine. Computer Networks and ISDN Systems 30(1-7), 107–117 (1998), http://linkinghub.elsevier.com/retrieve/pii/S016975529800110X
7. Burch, M., Vehlow, C., Beck, F., Diehl, S., Weiskopf, D.: Parallel edge splatting for scalable dynamic graph visualization. IEEE Transactions on Visualization and Computer Graphics 17(12), 2344–2353 (2011)
8. Clauset, A., Newman, M.E.J., , Moore, C.: Finding community structure in very large networks. Physical Review E, 1–6 (2004), http://www.ece.unm.edu/ifis/papers/community-moore.pdf
9. Correa, C.D., Crnovrsanin, T., Ma, K.L.: Visual reasoning about social networks using centrality sensitivity. IEEE Transactions on Visualization & Computer Graphics 18(1), 106–120 (2012)
10. Cox, T., Cox, M.: Multidimensional Scaling. Chapman & Hall, London (2001)
11. Crnovrsanin, T., Liao, I., Wuy, Y., Ma, K.L.: Visual recommendations for network navigation. In: Proceedings of the 13th Eurographics / IEEE – VGTC conference on Visualization, EuroVis 2011, pp. 1081–1090. Eurographics Association, Aire-la-Ville (2011), http://dx.doi.org/10.1111/j.1467-8659.2011.01957.x
12. Crnovrsanin, T., Muelder, C.W., Faris, R., Felmle, D., Ma, K.L.: Visualization of friendship and aggression networks (2012), http://vidi.cs.ucdavis.edu/projects/AggressionNetworks/, CNN's AC360 study: Schoolyard bullies not just preying on the weak
13. Demoll, B.S., Mcfarland, D.: The Art and Science of Dynamic Network Visualization. JoSS: Journal of Social Structure 7 (2005), http://www.cmu.edu/joss/content/articles/volume7/deMollMcFarland/
14. Dwyer, T., Koren, Y.: Dig-cola: Directed graph layout through constrained energy minimization. In: IEEE Symposium on Information Visualization, pp. 65–72 (2005)
15. Eades, P.: A Heuristic for Graph Drawing. Congressus Numerantium 42, 149–160 (1984)
16. Elmqvist, N., Fekete, J.D.: Hierarchical Aggregation for Information Visualization: Overview, Techniques, and Design Guidelines. IEEE TVCG 16(3), 439–454 (2009)
17. Faust, K.: Triadic configurations in limited choice sociometric networks: Empirical and theoretical results. Social Networks 30, 273–282 (2008)

18. Freeman, L.: Centrality in social networks conceptual clarification. Social Networks 1(3), 215–239 (1979)
19. Freeman, L.C.: The Development of Social Network Analysis: A Study in the Sociology of Science. Booksurge (2004)
20. Furnas, G.W.: Generalized fisheye views. In: Human Factors in Computing Systems CHI, pp. 16–23 (1986)
21. Golbeck, J., Robles, C., Edmondson, M., Turner, K.: Predicting personality from twitter. In: Proc. SocialCom 2011, pp. 149–156 (2011)
22. Gou, L., Zhang, X.: TreeNetViz: revealing patterns of networks over tree structures. IEEE Transactions on Visualization and Computer Graphics 17(12), 2449–2458 (2011)
23. Gou, L., Zhang, X., Luo, A., Anderson, P.: SocialNetSense: supporting sensemaking of social and structural features in networks with interactive visualization. In: 2012 IEEE Conference on Visual Analytics Science and Technology (VAST 2012), pp. 133–142 (2012)
24. Hachul, S., Jünger, M.: An experimental comparison of fast algorithms for drawing general large graphs. In: Healy, P., Nikolov, N.S. (eds.) GD 2005. LNCS, vol. 3843, pp. 235–250. Springer, Heidelberg (2006)
25. van Ham, F., Perer, A.: Search, Show Context, Expand on Demand: Supporting Large Graph Exploration with Degree-of-Interest. IEEE TVCG 15(6), 953–960 (2009)
26. Hasan, M.A., Zaki, M.J.: A survey of link prediction in social networks. In: Aggarwal, C.C. (ed.) Social Network Data Analytics, pp. 243–275. Springer US (2011)
27. Heer, J., Boyd, D.: Vizster: visualizing online social networks. In: IEEE Symposium on Information Visualization, pp. 32–39 (2005)
28. Holten, D.: Hierarchical edge bundles: Visualization of adjacency relations in hierarchical data. IEEE Transactions on Visualization and Computer Graphics 12(5), 741–748 (2006)
29. Holten, D.: Hierarchical edge bundles: Visualization of adjacency relations in hierarchical data. IEEE Trans. Vis. Comput. Graph. 12(5), 741–748 (2006)
30. Hu, Y., Kobourov, S.G., Veeramoni, S.: Embedding, clustering and coloring for dynamic maps. In: Proceedings of the 5th IEEE Pacific Visualization Symposium, pp. 33–40 (2012)
31. Huang, M.L., Nguyen, Q.V.: A fast algorithm for balanced graph clustering. In: Proceedings of the 2007 IEEE Symposium on Information Visualization (InfoVis), pp. 46–52 (2007)
32. Jacob, R., Koschützki, D., Lehmann, K., Peeters, L., Tenfelde-Podehl, D.: Algorithms for centrality indices. In: Brandes, U., Erlebach, T. (eds.) Network Analysis. LNCS, vol. 3418, pp. 62–82. Springer, Heidelberg (2005)
33. Kamada, T., Kawai, S.: An algorithm for drawing general undirected graphs. Inf. Process. Lett. 31(1), 7–15 (1989)
34. Kilduff, M., Tsai, W.: Social Networks and Organizations. SAGE (September 2003)
35. Kleinberg, J.M.: Authoritative sources in a hyperlinked environment. Journal of the ACM 46(5), 604–632 (1999), http://portal.acm.org/citation.cfm?doid=324133.324140
36. Langevin, D.G.S., Schretlen, P., Jonker, D., Bozowsky, N., Wright, W.: Louvain clustering for big data graph visual analytics (2013), poster at VIS 2013

37. Linden, G., Smith, B., York, J.: Amazon.com Recommendations: Item-to-Item Collaborative Filtering. IEEE Internet Computing 7, 76–80 (2003)
38. Lister, R.: After the gold rush: toward sustainable scholarship in computing. In: Simon, M., Hamilton (eds.) Tenth Australasian Computing Education Conference (ACE 2008). CRPIT, vol. 78, pp. 3–18. ACS, Wollongong (2008)
39. Mahmud, J., Zhou, M., Megiddo, N., Nichols, J., Drews, C.: Recommending targeted strangers from whom to solicit information on social media. In: Proc. IUI 2013, pp. 37–48 (2013)
40. Moscovich, T., Chevalier, F., Henry, N., Pietriga, E., Fekete, J.-D.: Topology-Aware Navigation in Large Networks. In: SIGCHI Conference on Human Factors in Computing Systems, pp. 2319–2328 (2009), http://hal.inria.fr/inria-00373679
41. Muelder, C., Ma, K.L.: A treemap based method for rapid layout of large graphs. In: Proceedings of the IEEE Pacific Visualization Symposium (PacificVis 2008), pp. 231–238 (2008)
42. Muelder, C., Ma, K.L.: Rapid graph layout using space filling curves. IEEE Transactions on Visualization and Computer Graphics 14(6), 1301–1308 (2008)
43. Muelder, C.W., Crnovrsanin, T., Ma, K.L.: Egocentric storylines for visual analysis of large dynamic graphs. In: Proceedings of 1st IEEE Workshop on Big Data Visualization (BigDataVis 2013), pp. 56–62 (October 2013)
44. Newman, M.E.J.: The Structure and Function of Complex Networks. SIAM Review 45(2), 167–256 (2003)
45. Noack, A.: Modularity clustering is force-directed layout. CoRR abs/0807.4052 (2008)
46. Pal, A., Wang, F., Zhou, M., Nichols, J., Smith, B.: Question routing to user communities. In: CIKM 2013 (to appear, 2013)
47. Pennacchiotti, M., Popescu, A.M.: A machine learning approach to twitter user classification. In: ICWSM (2011)
48. Qian, T., Li, Q., Liu, B., Xiong, H., Srivastava, J., Sheu, P.: Topic formation and development: a core-group evolving process. In: WWW 2013, pp. 1–31 (2013)
49. Rivera, M.T., Soderstrom, S.B., Uzzi, B.: Dynamics of dyads in social networks: Assortative, relational, and proximity mechanisms. Annual Review of Sociology 36, 91–115 (2010)
50. Russell, D.M., Stefik, M.J., Pirolli, P., Card, S.K.: The cost structure of sensemaking. In: Proceedings of the INTERACT 1993 and CHI 1993 Conference on Human Factors in Computing Systems, CHI 1993, pp. 269–276. ACM, New York (1993), http://doi.acm.org/10.1145/169059.169209
51. Sallaberry, A., Muelder, C., Ma, K.-L.: Clustering, visualizing, and navigating for large dynamic graphs. In: Didimo, W., Patrignani, M. (eds.) GD 2012. LNCS, vol. 7704, pp. 487–498. Springer, Heidelberg (2013)
52. Stasko, J., Zhang, E.: Focus+context display and navigation techniques for enhancing radial, space-filling hierarchy visualizations. In: IEEE Symposium on Information Visualization, InfoVis 2000. pp. 57–65 (2000)
53. Sugiyama, K., Tagawa, S., Toda, M.: Methods for visual understanding of hierarchical systems. IEEE Trans. Systems, Man, and Cybernetics 11, 109–125 (1981)
54. Tanahashi, Y., Ma, K.L.: Design considerations for optimizing storyline visualizations. IEEE TVCG 18(12), 2679–2688 (2012)

55. Tollis, I.G., Di Battista, G., Eades, P., Tamassia, R.: Graph Drawing: Algorithms for the Visualization of Graphs. Prentice Hall (July 1999)
56. Tufte, E.R.: Envisionning Information. Graphics Press (1990)
57. White, S., Smyth, P.: Algorithms for estimating relative importance in networks. In: Proceedings of the 9th ACM SIGKDD International Conference on Knowledge Discovery and Data Mining, KDD, pp. 266–275 (2003)
58. Zhao, S., Zhou, M., Zhang, X., Yuan, Q., Zheng, W., Fu, R.: Who is doing what and when: Social map-based recommendation for content-centric social web sites. ACM TIST 3(1), 5–25 (2011)
59. Zhou, M., Zhang, W., Smith, B., Varga, E., Farias, M., Badenes, H.: Finding someone in my social directory whom i do not fully remember or barely know. In: Proc. ACM IUI 2012, pp. 203–206 (2012)

4
Multivariate Networks in the Life Sciences

Oliver Kohlbacher, Falk Schreiber, and Matthew O. Ward

Data in the life sciences is being obtained at a steadily increasing speed. Modern technology enables observing many of the fundamental building blocks of a cell such as genes and their activity or metabolites and their concentration, as well as many phenotypical parameters on a macroscopic level, such as shape, volume or tissue composition. The sequencing of a large number of genomes—the blueprints of life—enabled so-called post-genomics methods. The suffix '-omics' indicates the generation of data on a large, comprehensive scale. Genomics thus studies all genes and proteomics all proteins in a cell or a tissue. Recent developments have led to a staggering list of these omics technologies. Some of the more popular omics technologies and the data associated with them include:

- *Genomics:* DNA sequence and genes
- *Transcriptomics:* mRNA sequence and expression levels
- *Proteomics:* protein sequence and expression levels
- *Metabolomics:* metabolite concentrations
- *Interactomics:* protein-protein interactions

Each of these data types requires different technologies for its generation. In genomics, DNA is extracted and fragmented into a library of small segments that are each sequenced in parallel. These sequence reads are then reassembled and annotated to derive genes. In transcriptomics, sample RNA is extracted and amplified. The expression level of each mRNA can then be estimated by next-generation sequencing (RNA-Seq) or by hybridization to oligonucleotide probes (microarrays). The key technology in proteomics and metabolomics is currently mass spectrometry, where peptides (derived from proteins by enzymatic digestion) or metabolites are separated by chromatographic techniques and then detected in a high-resolution mass spectrometer. The resulting datasets of most of these technologies are huge (up to terabytes per sample) and often extremely complex.

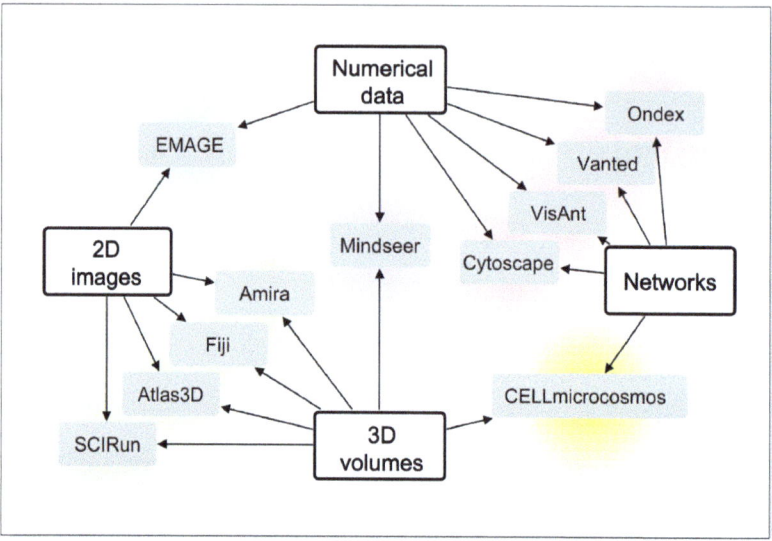

Fig. 4.1. Major data types in the life sciences and some bioinformatics tools that integrate more than one data type into the analysis process, see [17] for details

Several other types of data and information can also be integrated into the analysis process, including images, volumes, and text documents. Figure 4.1 shows a sampling of bioinformatics tools that integrate multiple forms of data.

4.1 Characteristics of Data and Tasks

Depending on the application, the data sources, and the questions under investigation, the resulting multivariate graphs can be very different. They will differ both in the semantics of the nodes and edges[1] (the type of the network) as well as in the data attached to nodes and edges.

4.1.1 Types of Biological Networks

Omics data is characterized as high-throughput and high-dimensional, as many parameters are measured at once. It is often of limited accuracy, as much noise exists in the process of extracting the data. Finally, the analysis can be quite complex, drawing from techniques found in statistics, data mining, machine learning, pattern recognition, as well as visualization.

While networks have been used for biological visualization for a long time (e. g., phylogenetic trees have been used since the early 1800s), the availability

---

[1] *Nodes* are also called *vertices* and *edges* are called *links*, respectively.

of high-throughput data resulted in network data on an unprecedented scale. This gave rise to the idea of 'network biology', understanding biology in terms of networks [3].

Omics data in the life sciences either represents a network (e. g., interactomics or regulomics) or can be interpreted in the context of a network (e. g., proteomics, transcriptomics, and metabolomics). Analysts may study these networks in many ways. They may focus just on a single network or part of a network, they may be interested in the interconnection between different networks, or they may want to compare multiple networks at once. In addition, they may wish to project a wide range of different data onto the networks, either on the nodes or the links, which is why the development of visualization techniques for multivariate networks is so important.

Biological networks can be organized into a hierarchy based on the entities represented by nodes and edges (see Fig. 4.2). From metabolic processes happening on an atomic scale to ecological and evolutionary networks taking place on planet-wide scales these networks cover a wide range of scales with respect to time and space. The networks differ mainly in the type of biological entities or processes represented by their nodes and edges:

- *Molecular graphs:* nodes are atoms, links are bonds.
- *Metabolic networks:* nodes are metabolites, links are reactions.
- *Interaction networks:* nodes are proteins, links are interactions.
- *Regulatory networks:* nodes are proteins, links are actions (activation, repression etc.).
- *Ecological networks:* nodes are species, links are interactions.
- *Evolutionary networks:* nodes are species, links indicate evolution.

This list is neither complete nor uniquely defined. Multiple representations are possible for many of these networks. The entities present in one network type (or layer) often have equivalents in other network types. A reaction node in a metabolic network represents an enzyme, which can interact with other proteins and is thus also represented by a node in an interaction network, or can be regulated by other genes or gene products (see Fig. 4.3).

Layouts can be either overlapping or non-overlapping. Nesting of nodes is possible to show hierarchical relationships. Additional marks and symbols can be incorporated to convey direction of relationships, locations within a cell or organism, and other types of meta-data.

4.1.2 Data Mapping and Multivariate Networks

The choice of networks underlying the data depends on the application and on the available data. In most cases, the structure of the networks is more or less fixed and the network data is taken from curated databases (such as KEGG [7], Reactome [13], BIND [2], and DIP [18]). This reflects the fact that within a given species, the structure of most networks shows little

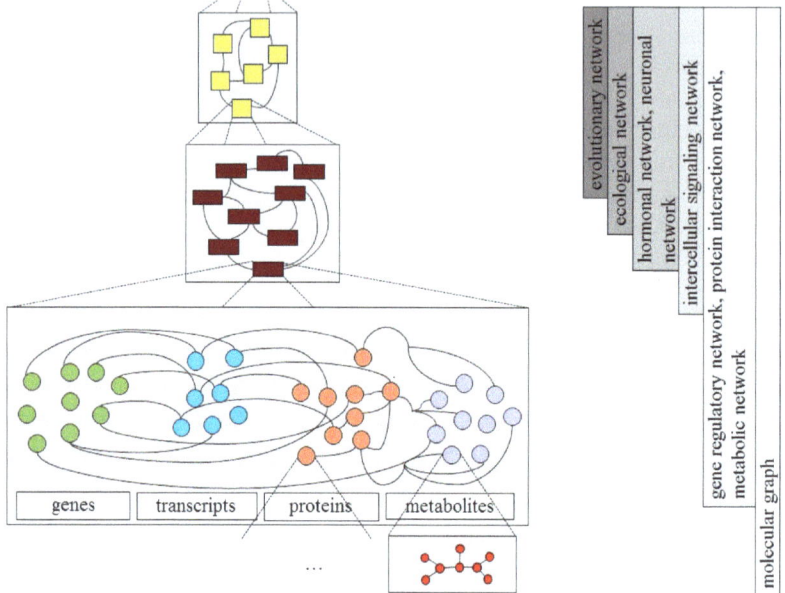

Fig. 4.2. A hierarchy of biological networks

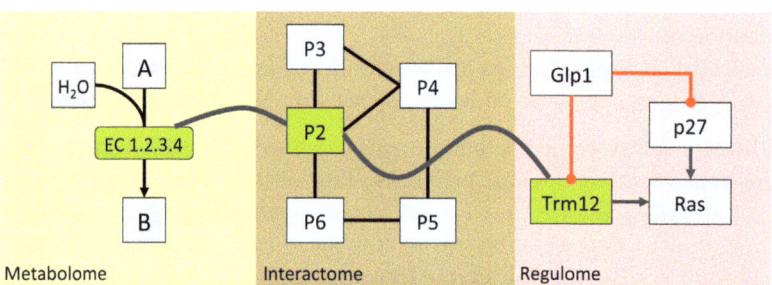

Fig. 4.3. Different levels of networks are connected through shared entities

variation[2]. What changes, though, is the state of the network, such as the concentrations of metabolites as a function of time or the expression level of genes as a function of the tissue.

The purpose of network visualization is thus, more often than not, to show the omics data in the context of these networks. Due to the size of the underlying networks, it is usually not meaningful to visualize the whole network. In most cases, only parts of the whole network are relevant and these can be identified by statistical means. For example, so-called enrichment analyses

---

[2] Although it should be noted that the network data itself is often incomplete, and therefore, the networks change over time due to increasing knowledge.

4.1 Characteristics of Data and Tasks   65

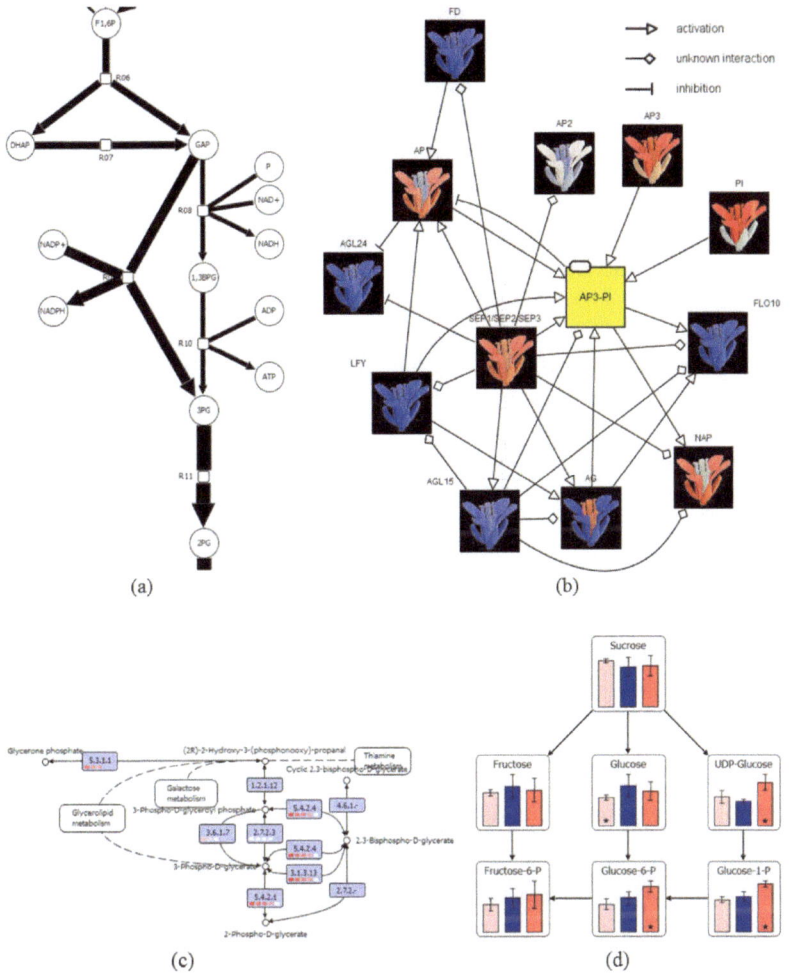

Fig. 4.4. Examples of multivariate data in biological networks, (a) flux data in metabolic networks [15], (b) spatial resolution of gene expression in a gene regulatory networks of Arabidopsis [6], (c) expression data mapped on a KEGG pathway [7], and (d) metabolite concentrations under three different conditions

can identify subnetworks that show statistically significant changes in expression levels [19]. The visualization can thus be focused onto the relevant parts of the network only, omitting the unchanged parts. In Fig. 4.4, as well as in Sect. 4.2, we give some examples of how these network and omics datasets can be represented. Typical graphical attributes used on nodes and edges to convey information are:

- *Nodes:* text labels, shape, size, color, diagrams, etc.
- *Edges:* text labels, line style, thickness, color, etc.

4.2 Use Cases

Here we discusses some use cases that show a variety of networks and ways in which multivariate biological network data has been visualized in the past.

4.2.1 Signaling

Signaling in cells can be conveyed via different mechanisms. One of the best-studied of these mechanisms is the chemical modification (phosphorylation) of certain amino acids of a protein (serine, threonine, tyrosin). This modification is reversible and is usually catalyzed by specific enzymes (kinases, phosphatases). By modifying amino acid sites in a protein very specifically, the activity of these proteins can be modulated – they can be activated or deactivated. If kinases or phosphatases themselves are activated or deactivated, they can in turn change the phosphorylation of other enzymes/proteins. In this way, a signal can be transmitted from one protein to another. This information flow follows well-defined *signaling pathways* and these pathways are part of large *signaling networks*. Signaling itself plays a key role in many biological processes and proteomics provides a time-resolved view of these signaling events. In order to unravel these networks, i. e., to figure out which protein activates which other protein at what timepoint, the visualization of these datasets in a larger context is quite helpful. In the example above, we visualized the phosphorylation patterns as a function of time (Fig. 4.5) for those nodes of the network for which (phospho-)proteomics could determine the phosphorylation patterns. Analysis of these patterns can be used to understand the dynamic behavior of signaling networks.

4.2.2 Genetic Linkage

Genetic linkage analysis is focused on the tendencies of genes that are close to each other on a chromosome to be inherited together during meiosis (cell division necessary for sexual reproduction in eukaryotes). A set of genes or gene markers undergo pairwise comparison to ascertain how frequently they undergo recombination during crossover of homologous chromosomes. This linkage score reflects the frequency of recombination between two markers or genes, which is an indication of their genetic distance (as well as physical distance). CheckMatrix (http://cgpdb.ucdavis.edu/XLinkage) is a visualization tool for analyzing and validating genetic maps. It uses a set of genetic markers (x and y axes in matrix) and recombination/linkage data for all possible pairs of markers computed via a variety of algorithms to create a matrix, where the color of each cell is based on the linkage score (see Fig. 4.6). Along the right border are the names for the markers and their positions in the sequence. Allele composition is shown along the bottom.

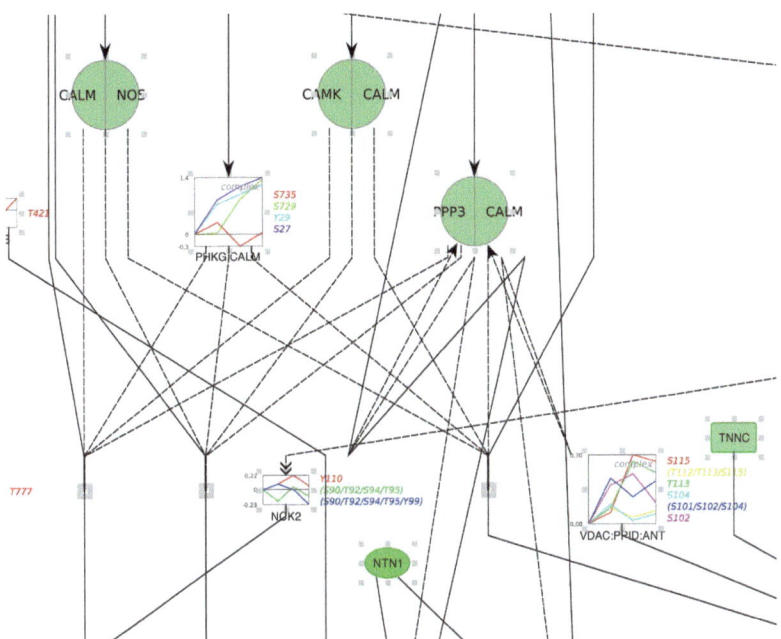

Fig. 4.5. Visualization of a signaling network showing timeseries data of the phosphorylation patterns of selected proteins. The different curves in each box represent different phosphorylation sites in the same protein.

4.2.3 Relationship Discovery Based on Document Analysis

While much biological data visualization is focused on the analysis of data sets containing sequences, numbers, and images, there is a growing interest in harvesting information from large document repositories such as PubMed (http://www.ncbi.nlm.nih.gov/pubmed). Chilibot (CHIp LIterature roBOT) is a tool that accepts a user's set of input keywords and gene symbols and mines PubMed abstracts for relations between the supplied terms [4]. It first augments the list with synonyms compiled from several databases (users can add to this table) and then does sophisticated natural language processing on each sentence of a collection of retrieved abstracts to find not only co-occurrences, but also types of relationships (stimulatory, inhibitory, neutral, parallel, and simple co-occurrence). The visualization represents query terms as boxes and relations as lines. Box colors are set based on degree of up/down regulation from experimental data, while line color is based on whether the relationship is stimulatory (green), inhibitory (red), or both. Grey lines are neutral. Each edge also can have a circled number indicating how many abstracts contained information about the relationship. Mousing over an edge or node provides a text annotation of the relationship or term extracted from the abstracts. Finally, arrows are added if the abstract

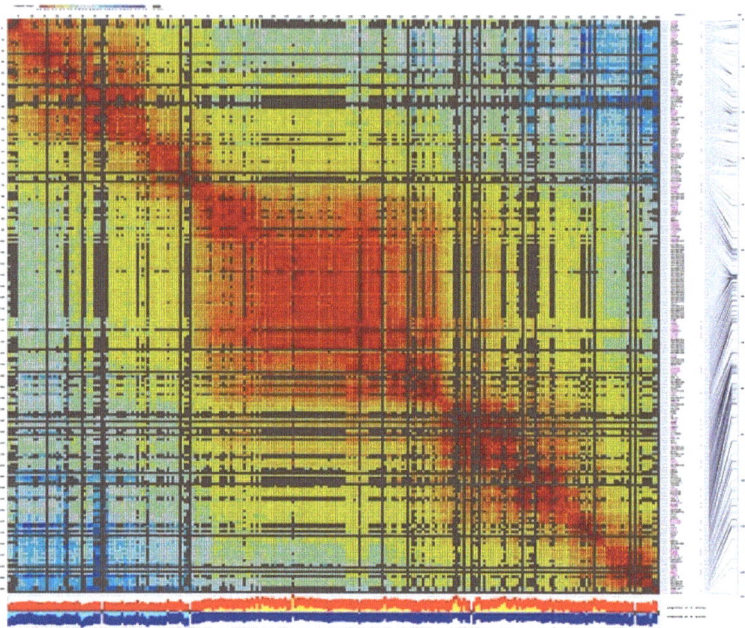

Fig. 4.6. A transcript-based genetic map generated by CheckMatrix, showing linkage information for a set of genetic markers from chromosome 4 from the plant Arabidopsis. Python_MadMapper BIT scores are mapped to color [20].

indicated directionality of the relationship. Grey diamonds indicate only a co-occurrence relationship exists. See Fig. 4.7 for an example application.

4.2.4 Gene Regulation and Transcriptome Data

Gene regulation is a complex process commonly represented by gene regulatory networks. Both the static structure of the network as well as the dynamics of regulatory events are important to understand gene regulation. The static structure of a gene regulatory network is often used to investigate functional building blocks derived from network motifs [14] or central regulatory nodes based on network centrality analysis [11]. Dynamic changes, such as organ development and morphological characteristics of higher organisms, can be traced back to gene regulatory events, which are shown by changes in the expression level of genes. The steadily increasing temporal and spatial resolution of transcriptome datasets (measuring the expression levels of genes) requires a set of analysis methods including exploration and visualization to provide insights into developmental processes.

An example is shown in Fig. 4.4(b), where we consider the visualization and exploration of tissue-specific gene expression data for master regulators of

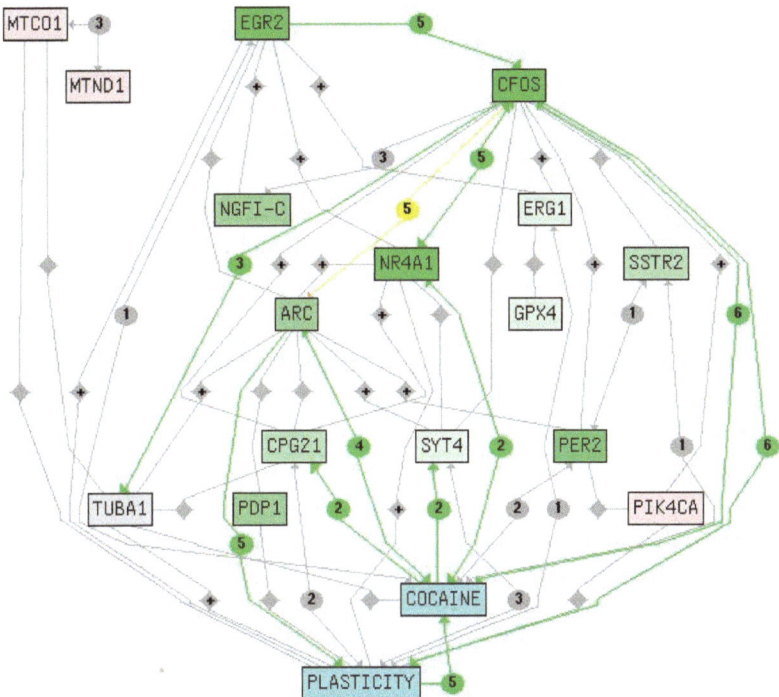

Fig. 4.7. Using Chilibot to study the relationships between genes reported to be regulated by cocaine. The network is formed automatically based on discovered relations [4].

Arabidopsis thaliana flower development in the context of the corresponding gene regulatory network [6]. In the network in Fig. 4.4(b), nodes represent genes and different types of links (represented by different arrow heads) are used to represent information about activation and inhibition. The nodes contain color-coded images which show the expression levels of the genes represented by the network node in different floral organs of the plant *Arabidopsis*.

The combination of network, omics data, and spatial information provides a fast visual exploration not only of regulatory events, but also of similar and different expression of a specific gene in the context of different tissues or organs (spatial context). Such representations can support the comparative analysis of genes with specific transcript patterns, thereby helping in extracting functional relationships.

4.3 Challenges

High-throughput data is rapidly growing in popularity in all areas of research in the life sciences. This implies that more and more non-experts get in contact with this type of data and are forced to tackle the complexity of analyzing complex multi-omics data sets. Further background information concerning the interactive visual analysis of biological networks (in particular information visualization, visual analytics, and automatic layout of networks) is given in [9]. Although there are many tools available for biological network visualization (for overviews and comparisons see, for example, [5, 10, 16]), there are still many challenges to be met [1]. The challenges arise partially from the growing amount of available high-throughput data, partially from novel applications, partially from the integration of different networks, and partially from the increasing need of more user-friendly visual analytics tools.

Currently, the key challenges concern *scale, uncertainty/ambiguity, heterogeneity, interactivity* and *standardization*. We will discuss each of these challenges separately in the following.

4.3.1 Scale

For some biological processes the complete networks have to be taken into consideration and thus need to be visualized. Currently networks range from a few dozens to a few thousand nodes and up to several thousand edges (for example, protein interaction or whole-genome metabolic networks), and networks with hundreds of nodes and thousands of links are in common use. This likely will expand by at least one order of magnitude in the near future. So far, tools commonly lack good methods to navigate through such large networks.

In addition, the amount and complexity of multivariate date (especially omics data, but also images, volumes, texts and so on) is steadily increasing. To make sense out of the data their integration into cellular processes and biological networks is often required. This also has implications for interactivity, exploration, and visualization. See Chap. 10 for scalability considerations for multivariate graph visualization.

4.3.2 Uncertainty/Ambiguity

Unlike in some domains, relations and values in bioinformatics are never one hundred percent certain. Concerning the structure of the networks, generally there is evidence to support a relationship, but it could be a very weak correlation that may, as more evidence is analyzed, prove to be incorrect. Also the data mapped onto the networks is often uncertain. Both the uncertainty of the network structure (and thereby the reliability of the underlying network data) as well as the uncertainty of the different related data has to be shown to a user.

Typical examples are measurement errors, missing data, multiple solutions produced by algorithms (such as in the process of finding mappings from one sequence to another, most search algorithms will report only the best match found, but in reality there may be multiple matches for the same subsequence of comparable quality), and ambiguous mappings between elements of different domains.

4.3.3 Heterogeneity

While most multivariate network visualizations incorporate a single data type, it is increasingly important to tie different data types within the analysis process. The result are heterogeneous networks with a complex structure: different types of nodes, edges, hyper edges, and hierarchical relationships (see Fig. 4.2).

Two major challenges are: (1) the compilation of heterogeneous networks which requires the identification of the biological entities and the interconnection between networks. The interconnection is especially difficult to obtain as identifiers for biological entities are often only unique in the context of one data source, for example, a database or an ontology, and identifier mapping mechanisms have to be established. (2) the visualization and interactive exploration of heterogeneous networks, which so far has not been sufficiently solved. See Chap. 9 for discussions of heterogeneous networks at multiple levels.

4.3.4 Interactivity

The scale and complexity of the data implies that discovery of new biological insights requires large-scale data exploration. Network visualization is thus more and more tied into visual analytics workflows [8]. In order to make such tools usable, interactive response times, mental map preserving animations, and easy to use interfaces are required to achieve acceptance in the user base. See Chap. 6 for discussion of interaction in the visualization of multivariate networks.

4.3.5 Standardization

Standardized glyphs for different node and link types are common in other areas of science, such as electrical engineering. Such uniform systems of nomenclature that describe the components of networks and are based on a well-defined set of symbols greatly facilitates communication efficiency and clarity. Although many visualizations in biology still do not follow uniform rules, graphical standards such as the Systems Biology Graphical Notation [12] have been established and should be obeyed to foster better understanding of network visualizations in biology.

4.4 Summary and Conclusions

In this chapter, we have described the broad range of biological data that is being routinely collected and analyzed, ranging from the atomic to the planetary scale. Data is not only available in the form of genetic sequences and numeric tables, but also in the form of images, volumes, text, and relational information. This relational information, whether explicit in the data or implicitly derived, is then the focus of multivariate network visualization. We then briefly described the typical mappings of such data to networks and presented a number of case studies showing their use in performing a variety of bioinformatics tasks. Finally, we concluded with our views on some of the key challenges facing the field of network visualization in bioinformatics.

In the future, we expect that visualization and interactive exploration will play an increasingly important role in the study of biological data and processes. This will lead to not only increased understanding of how living organisms develop, but also their relationships to other organisms. It will also be a key factor in expanding our understanding of diseases and lead to improved methods for their treatment. As mentioned in Sect. 4.3, there are many challenges that will need to be overcome in order to achieve these goals. We expect that new biological data types, as well as increased needs to integrate these types into the analytics process, will provide a wealth of opportunities for visualization researchers for many years to come.

References

1. Albrecht, M., Kerren, A., Klein, K., Kohlbacher, O., Mutzel, P., Paul, W., Schreiber, F., Wybrow, M.: On open problems in biological network visualization. In: Eppstein, D., Gansner, E.R. (eds.) GD 2009. LNCS, vol. 5849, pp. 256–267. Springer, Heidelberg (2010)
2. Bader, G.D., Betel, D., Hogue, C.W.: BIND: the biomolecular interaction network database. Nucleic Acids Research 31(1), 248–250 (2003)
3. Barabasi, A.L., Oltvai, Z.N.: Network biology: understanding the cell's functional organization. Nature Reviews Genetics 5(2), 101–113 (2004)
4. Chen, H., Sharp, B.M.: Content-rich biological network constructed by mining PubMed abstracts. BMC Bioinformatics 5(1), 147 (2004)
5. Gehlenborg, N., O'Donoghue, S.I., Baliga, N.S., Goesmann, A., Hibbs, M.A., Kitano, H., Kohlbacher, O., Neuweger, H., Schneider, R., Tenenbaum, D., Gavin, A.C.: Visualization of omics data for systems biology. Nature Methods 7, S56–S68 (2010)
6. Junker, A., Rohn, H., Schreiber, F.: Visual analysis of transcriptome data in the context of anatomical structures and biological networks. Frontiers in Plant Science 3, 252 (2012)
7. Kanehisa, M., Goto, S., Hattori, M., Aoki-Kinoshita, K.F., Itoh, M., Kawashima, S., Katayama, T., Araki, M., Hirakawa, M.: From genomics to chemical genomics: new developments in KEGG. Nucleic Acids Research 34, D354–D357 (2006)

8. Kerren, A., Schreiber, F.: Toward the role of interaction in visual analytics. In: Rose, O., Uhrmacher, A.M. (eds.) Proceedings of the Winter Simulation Conference (WSC 2012). pp. 420:1–420:13 (2012)
9. Kerren, A., Schreiber, F.: Network visualization for integrative bioinformatics. In: Approaches in Integrative Bioinformatics: Towards the Virtual Cell, pp. 173–202. Springer (2014)
10. Kono, N., Arakawa, K., Ogawa, R., Kido, N., Oshita, K., Ikegami, K., Tamaki, S., Tomita, M.: Pathway projector: Web-based zoomable pathway browser using KEGG atlas and Google maps API. PLoS One 4(11), e7710 (2009)
11. Koschützki, D.: Network centralities. In: Junker, B.H., Schreiber, F. (eds.) Analysis of Biological Networks. Wiley Series on Bioinformatics, Computational Techniques and Engineering, pp. 65–84. Wiley (2008)
12. Le Novère, N., Hucka, M., Mi, H., Moodie, S., Schreiber, F., Sorokin, A., Demir, E., Wegner, K., Aladjem, M., Wimalaratne, S.M., Bergman, F.T., Gauges, R., Ghazal, P., Kawaji, H., Li, L., Matsuoka, Y., Villéger, A., Boyd, S.E., Calzone, L., Courtot, M., Dogrusoz, U., Freeman, T., Funahashi, A., Ghosh, S., Jouraku, A., Kim, S., Kolpakov, F., Luna, A., Sahle, S., Schmidt, E., Watterson, S., Wu, G., Goryanin, I., Kell, D.B., Sander, C., Sauro, H., Snoep, J.L., Kohn, K., Kitano, H.: The systems biology graphical notation. Nature Biotechnology 27, 735–741 (2009)
13. Matthews, L., Gopinath, G., Gillespie, M., Caudy, M., Croft, D., de Bono, B., Garapati, P., Hemish, J., Hermjakob, H., Jassal, B., Kanapin, A., Lewis, S., Mahajan, S., May, B., Schmidt, E., Vastrik, I., Wu, G., Birney, E., Stein, L., D'Eustachio, P.: Reactome knowledgebase of human biological pathways and processes. Nucleic Acids Research 37(1), D619–D622 (2009)
14. Milo, R., Shen-Orr, S., Itzkovitz, S., Kashtan, N., Chklovskii, D., Alon, U.: Network motifs: Simple building blocks of complex networks. Science 298(5594), 824–827 (2002)
15. Rohn, H., Hartmann, A., Junker, A., Junker, B.H., Schreiber, F.: FluxMap: a Vanted add-on for the visual exploration of flux distributions in biological networks. BMC Systems Biology 6, 33 (2012)
16. Rohn, H., Junker, A., Hartmann, A., Grafahrend-Belau, E., Treutler, H., Klapperstck, M., Czauderna, T., Klukas, C., Schreiber, F.: VANTED v2: a framework for systems biology applications. BMC Systems Biology 6(139) (2012)
17. Rohn, H., Klukas, C., Schreiber, F.: Creating views on integrated multidomain data. Bioinformatics 27(13), 1839–1845 (2011)
18. Salwinski, L., Miller, C.S., Smith, A.J., Pettit, F.K., Bowie, J.U., Eisenberg, D.: The database of interacting proteins: 2004 update. Nucleic Acids Research 32(1), 449–451 (2004)
19. Subramanian, A., Tamayo, P., Mootha, V.K., Mukherjee, S., Ebert, B.L., Gillette, M.A., Paulovich, A., Pomeroy, S.L., Golub, T.R., Lander, E.S., Mesirov, J.P.: Gene set enrichment analysis: A knowledge-based approach for interpreting genome-wide expression profiles. Proceedings of the National Academy of Sciences of the United States of America 102(43), 15545–15550 (2005)
20. Truco, M.J., Ashrafi, H., Kozik, A., van Leeuwen, H., Bowers, J., Wo, S.R.C., Stoffel, K., Xu, H., Hill, T., Van Deynze, A., et al.: An ultra-high-density, transcript-based, genetic map of lettuce. G3: Genes— Genomes— Genetics 3(4), 617–631 (2013)

Part II

Topics in Multivariate Network Research

5

Tasks for Multivariate Network Analysis

A. Johannes Pretorius, Helen C. Purchase, and John T. Stasko

In Chap. 1, a multivariate network was defined as having two important characteristics. First, nodes are connected to each other via links; there is topological structure. Second, being multivariate, nodes and links have attributes associated with them, with these attributes having a value.

In this chapter, we describe tasks associated with multivariate networks. We consider a task to be an activity that a user wishes to accomplish by interacting with a visual representation of a multivariate network. This implies that there is user intent [13], and that the network has been presented visually. At the highest level, this intent is usually described as the goal of obtaining *insight* about the data being studied [6].

Pragmatically, the notion of gaining insight from visualizations can be described as one or more very high-level tasks. As Amar and Stasko put it, tasks that "real people want to accomplish" [3]. These include:

- Make complex decisions, especially under uncertainty;
- Learn a domain;
- Identify the nature of trends;
- Predict the future;
- Identify the domain parameters;
- Discover correlative models;
- Formulate and verify hypotheses;
- Identify the effect of data uncertainties; and
- Identify sources of causation.

In the spirit of Amar and Stasko's work, we note that this is a sample of high-level tasks and not a complete list. These tasks are not specific to multivariate networks and are biased towards exploration and confirmation. We recognize that some users may have additional objectives, such as the presentation of data, which fall outside the scope of this chapter. However, in a context where achieving insight is the main driver and where multivariate networks are of interest, performing a task such as those listed above involves one or more of the following activities [8]: gain an understanding of the structural properties of the network;

find patterns, clusters, and correlations between several attributes of the nodes and links; and relate understanding about attributes and structure.

In this chapter, we describe in more detail how this is accomplished by presenting a framework of tasks for multivariate networks. Our objective is to present, to a general audience, a frame-of-reference that encapsulates the types of tasks typically encountered when analyzing multivariate network data. As a result, the work presented here is deliberately not overly theoretical or abstract. We first recap the entities and properties of multivariate networks. We then describe a general taxonomy for visualization tasks. Next, we introduce a framework for multivariate network tasks and show how these are composed of lower-level tasks of the general taxonomy. We follow with a short discussion before concluding.

5.1 Entities and Properties

In the abstract, a task involves performing an analytic activity on a combination of an *entity* (the "thing" that is being studied), and a *property* of that entity [1]. We note that different terminology is sometimes used; for example, some authors refer to entities as data cases, and to properties as attributes [2]. Notwithstanding, a task can be represented as a process [1]:

Select entity → Select property → Perform analytic activity.

There is typically a high degree of iteration; based on the outcome of the analytic activity, the user may wish to select another entity and/or property to analyze. When considering multivariate networks, the entities that users study are [5]:

- *Nodes;*
- *Links;*
- *Paths,* or sequences of nodes and links; and
- *Networks,* since users may want to include several networks in their analysis.

Multivariate networks also have two types of associated properties [1]:

- *Structural properties,* sometimes referred to as topology; and
- *Attributes,* associated with nodes and links.

To make the above more concrete, we briefly revisit examples of multivariate networks from the three application areas discussed in Chaps. 2, 3, and 4. In software engineering, analysts study entities including software packages, classes, and methods. Tasks include studying the links, such as method calls, between entities. Properties of nodes and links model features that are fundamental to understanding software including package, class, and method names, and method call durations. Multivariate networks in biomedicine include metabolic networks (nodes represent atoms, links represent bonds), interaction networks (nodes represent metabolites, links represent interactions), and regulatory networks (nodes represent proteins, links represent actions).

Again, properties are important to facilitate insight, for example, whether actions in regulatory networks activate or repress protein production.

Social networks are perhaps more familiar to many readers (see Chap. 3). In such a network, nodes represent people and links represent the relationships between people. By analysing paths between nodes, it is possible to derive knowledge. For example, even if two people have no direct relationship, if they both have a relationship with a third person, there exists an indirect relationship between them that may (or may not) be of interest. For social network analysis, there are scenarios where it is useful to compare networks themselves. For example, behavioral biologists may be interested in comparing social networks of humans with those of other primates to identify similarities and differences.

A lot can be learned from studying the properties of social networks. For example, it is possible to derive which of two people is likely to have the greater influence on others by considering, respectively, the number of relationships they have with other people. Properties provide important information, such as the type of relationships (friendship versus professional, for example) and demographics (first name, last name, occupation, and so forth).

Combinations of network entities and properties give rise to more complex concepts. For example, basic networks with structural properties only are less complex than networks with single node and link attributes (often referred to as labels) which, in turn, are less complex than multivariate networks where nodes and links can have multiple attributes. Increased complexity of networks results in increasingly complex analyses [9], and this impacts the complexity of tasks that are performed. In cases where users want to compare two or more networks, there is an additional level of complexity.

It is also possible to calculate derived entities and properties, that is, entities and properties that do not explicitly exist in the underlying data. Two common derived entities are clusters and groups. *Clusters* are regions of networks that are structurally highly connected (these are sometimes referred to as cliques, particularly in social network analysis). *Groups* are subsets of nodes and links that share similar attribute properties. Examples of derived properties include statistical measures computed for a particular attribute (mean, median, and so forth).

Derived entities and properties are often involved in multivariate network tasks. As suggested, in social network analysis, clusters indicate cliques, or collections of people who have a high degree of interaction. Grouping could be used, for example, to identify and compare sets of people with similar demographics.

5.2 Tasks

As highlighted above, tasks involve entities (nodes, links, paths, networks) and properties of those entities (structural and attributes). The third component that makes up a task is the analytic activity, or the analysis. Below, we deconstruct tasks by focusing on different levels of analytic activity. Throughout this, we also refer to the entities and properties that are involved in tasks in a multivariate network context.

We first outline a general taxonomy for interactive visualization and then describe how some of these tasks are combined to form more complex tasks specific to multivariate networks.

5.2.1 General Task Taxonomy

Many authors have proposed general task taxonomies for information visualization. In seminal work, Wehrend and Lewis propose a classification of visualization methods by considering the entities being studied and tasks performed on the entities [12]. Specifically, they list 11 tasks that are frequently encountered:

- Identify;
- Locate;
- Distinguish;
- Categorize;
- Cluster;
- (Analyze) distribution;
- Rank;
- Compare;
- (Analyze) within and between relations;
- Associate; and
- Correlate.

By synthesizing questions that users typically have about their data, Amar et al. propose a different list of information visualization tasks [2]:

- Retrieve value;
- Filter;
- Compute derived value;
- Find extremum;
- Sort;
- Determine range;
- Characterize distribution;
- Find anomalies;
- Cluster; and
- Correlate.

In other related work on general information visualization taxonomies, Schulz et al. recently proposed a classification of the "design space" of visualization tasks based on five dimensions (*goal; means; characteristics,* or level of analysis; *target,* the parts of the data to be considered; and *cardinality,* the number of data instances to be considered) [7]. This allows a formal faceted specification of tasks by five-dimensional tuples. Brehmer and Munzner propose descriptions of visualization tasks by considering three aspects [4]: *Why* is the data being analyzed?; *How* is it being analyzed?; and *What* are the task inputs and outputs? In particular, they stress the difference between how (the means) and why (the goal) a task is performed. There are clear parallels between these approaches: why relates to goal; how relates to means; and what encapsulates characteristics, target, and cardinality.

These approaches are very general and abstract (Shulz et al. write that theirs is "applicable by a limited number of visualization experts only") and do not easily support the definition of a detailed taxonomy; we return to them in the conclusion. More pragmatically, it is worth noting the similarities between the sets of tasks proposed by Wehrend and Lewis [12] and Amar and Stasko [3], for example, both make provision for studying distributions. However, a like-for-like comparison is not immediately obvious. Further, it could also be argued that both Wehrend and Lewis's and Amar and Stasko's list of tasks operate at varying levels, for example, a task such as "filter" is more of an operational task while "correlate" is more of an analytical one.

The work by Valiati et al. addresses such difficulties by distinguishing three broad classes of tasks [11]: operational (relating to the means by which the network is presented and explored), analytical (the means by which information is extracted from the network), and cognitive (facilitating understanding of the whole network). Each category comprises one or more tasks. Accordingly, the taxonomy put forward in this chapter is based on the categories and constituent tasks defined by Valiati et al., classified as follows:

- *Operational:* visualize, configure;
- *Analytical:* identify, determine, relocate, compare; and
- *Cognitive:* infer.

While we acknowledge the fundamental facilitating role that operational tasks play in making the relevant information visible, most of our emphasis will be on the analytical category. Cognitive tasks are also considered, keeping in mind that the purpose of the whole exercise is, of course, to support the cognitive task of obtaining insight (as described by the high-level tasks of Amar and Stasko [3] and listed in the introduction). To avoid additional complexity, we reuse the terminology proposed by Valiati et al., which in turn, is based on the work by Wehrend and Lewis, although some terms could arguably be substituted with other descriptive verbs.

Operational Tasks

Operational tasks are concerned with the means of presenting the network to the user, and the facilities provided for the user to explore the data. These tasks are therefore more associated with the nature of the information visualization tool than with the user's tasks per se.

- *Visualize.* Invoke a particular graphical representation or a combination of graphical representations to visualize the entities and properties of a

multivariate network. The visualize task does not necessarily imply that all entities and all properties in the data are shown. In fact, it is almost always performed in combination with the configure task (described below) to selectively show or hide certain entities and/or properties.
- *Configure.* Interactively set up or change the visual representation in support of the analytical tasks. Typical visual configuration tasks include zoom, filter, and showing details on demand [8]. Much of the power of visualization, in general, is attributed to the combination of interactive configuration and corresponding real-time updates of the graphical representation [10].

Analytical Tasks

Analytical tasks are the primary building blocks for achieving a user's goal; they are the means by which specific information is obtained from the network. Analytical tasks are necessarily low-level, and applied to either individual entities or a small subset.

- *Identify.* Find entities and/or properties in the data. At an elementary level, the identify task involves discovery of entities based on their spatial location, or based on the values of associated properties as graphically encoded in one or more visual representations. In particular, the identify task often involves finding entities in networks that are adjacent with respect to the structure of the network. The identify task can also be more involved, however, and includes the visual identification of similarities, differences, patterns, outliers, variations, relationships (proximity, dependency, independency), and uncertainty.
- *Determine.* Calculate derived properties not originally present in the data. This often involves deriving statistical measures of the properties associated with nodes and links. Examples include: sum, difference, ratio, percentile, mean, median, variance, standard deviation, correlation coefficient, and probability. In addition, the determine task includes algorithmic calculation of derived entities, for example, clustering algorithms. As the result of invoking the determine task, the visualization is typically changed and, consequently, there is some overlap with the configure task, described above.
- *Relocate.* Revisit entities and/or properties already identified or determined. This implies that the user is already aware of the existence of these entities and/or properties, but has to exert effort to find them again. In some cases this may be trivial, but in others this may require as much effort as the initial identification task.
- *Compare.* Examine data entities and/or properties that have been identified and/or determined in contrast to each other. This often implicitly involves the relocate task. Comparisons are usually made to find similarities or differences between the properties of nodes and links. Because these properties are visually encoded, the compare task involves contrasting spatial location and/or graphical characteristics of the visual representation of the data.

Cognitive Tasks

- *Infer.* Derive insight or knowledge from the data as an outcome of performing a sequence of operational and analytical tasks. The range of outcomes of an infer task is broad: it may involve forming a hypothesis, or testing a hypothesis; it may be the result of explorative analysis or serendipitous discovery; it may lead to confirmation of an expectation, to insight that contradicts expectations, or to completely new knowledge. Such knowledge may take the form of cause-and-effect relationships, trends, or probabilities.

Cognitive tasks are high-level, relate specifically to "obtaining insight" [6], and are often iteratively developed by building on prior operational, analytical, and/or cognitive tasks. The Amar and Stasko tasks, which support users in achieving high-level objectives (as outlined in the introduction), are encompassed by this category [3]. Unlike analytical tasks, cognitive tasks are often associated with uncertainty and estimation. It is possible to determine whether the result of performing an analytical task has resulted in the "correct" answer. However, cognitive tasks are more complex and tend to require significant external resources (for example, memory storage, algorithms, or computational processing time) and the notion of "accuracy" does not exist for these tasks unless such support is provided. For this reason, the unsupported execution of a cognitive task may result in an uncertain or estimated answer.

5.2.2 Tasks for Multivariate Network Analysis

The task taxonomy introduced above is very general and can be applied to any data type. To meet the objective of this chapter (to describe tasks for multivariate networks), we now narrow the scope by introducing the network task taxonomy proposed by Lee et al. [5]. For network analysis, they propose four categories of tasks: *topology-based, attribute-based, browsing,* and *overview.* Lee et al.'s framework was devised by considering existing task taxonomies, by considering examples of tasks encountered in applications of network visualization to domain problems, and by reviewing tasks involved in user studies of network visualization methods.

Lee et al.'s framework is comprehensive in that it describes tasks commonly encountered when analyzing networks. To achieve this, they propose a number of tasks for each of the categories outlined above. However, these tasks are rather node-centric in the sense that nodes are generally assumed to be the entities of interest. Consequently, although we mirror quite closely the tasks proposed by Lee et al., we have generalized these to cater also for cases where other entities, such as links, or derived entities, such as clusters, are of interest to users. We also use slightly different terminology to that originally proposed. To avoid confusion with the more restrictive meaning of "topology" in a mathematical context, we refer to the first category of tasks as "structure-based". Also, we use the term "estimation tasks" as opposed to "overview tasks" as we find that the implied meaning more closely resembles the act of imprecisely or informally gauging general network characteristics.

The premise of Lee at al.'s work is that all tasks in the categories introduced above can be considered as conjunctions of general lower-level tasks. For this, they originally used the elementary tasks proposed by Amar et al. [2]. However, we employ the tasks described in the previous section, as proposed by Valiati et al. [11], because they address some of the shortcomings of other general task taxonomies (as highlighted in the previous section). We make one exception to the approach of composing network tasks from more general tasks, however. For estimation tasks, if a precise decomposition was possible, a "correct" answer would be guaranteed, which we will argue is not the case.

Structure-Based Tasks

Adjacency tasks combine analytical tasks (identify, determine, locate, and compare) to infer knowledge about the adjacency of entities. Two entities are adjacent if there exists a path of length at most one that connects them. In most situations, once an adjacent entity has been found, the user will proceed to study a property of that entity.

Task	Adjacency (entities)
Description	Find the set of entities adjacent to an entity.
Examples	Find the *first names* of the **persons** directly adjacent to a **person** with the *first name* "Adam" and *last name* "Smith". Find the *types* of **relationships** directly adjacent to a **person** with the *first name* "Adam" and *last name* "Smith".
Decomposition	Identify/relocate **entity** with *property* + identify/relocate adjacent **entity** (repeated) + identify/relocate *property* of **entity** (optional).

Task	Adjacency (derived property)
Description	Find a derived property of the entities adjacent to an entity.
Examples	Find the *number* of **persons** adjacent to a **person** with the *first name* "Adam" and *last name* "Smith". Find the *number* of **relationships** of *type* "professional" to a **person** with the *first name* "Adam" and *last name* "Smith".
Decomposition	Identify/relocate **entity** with *property* + identify/relocate adjacent **entity** (repeated) + determine *derived property* of **entity**.

Task	Adjacency (extreme properties)
Description	Find the entity with the maximum/minimum number of adjacent entities.
Example	Find the *first name* and *last name* of the **person** with the most **relationships** of *type* "friendship".
Decomposition	Identify/relocate **entity** with *properties* + identify/relocate adjacent **entity** + determine *derived property* of **entity** + compare *property* of **entity**.

5.2 Tasks

Accessibility tasks combine analytical tasks (identify, determine, locate, and compare) to infer knowledge about the accessibility of entities. An entity is accessible from another entity if there exists a path of any length that connects them. In most situations, once an accessible entity has been found, the user will proceed to study a property of that entity.

Task	*Accessibility (entities)*
Description	Find the set of entities accessible from an entity.
Example	Find the *first names* and *last names* of the friends of friends of a **person** with the *first name* "Adam" and *last name* "Smith". Find the *types* of **relationships** of the friends of friends of a **person** with the *first name* "Adam" and *last name* "Smith".
Decomposition	Identify/relocate **entity** with *property* + identify/relocate adjacent **entity** (repeated) + identify/relocate *property* of **entity** (optional).

Task	*Accessibility (derived properties)*
Description	Find a derived property of entities accessible from an entity.
Example	Find the *number* of **persons** with direct or indirect *relationships* of **type** "managed by" to a person with the *first name* "Adam" and *last name* "Smith".
Decomposition	Identify/relocate **entity** with *property* + identify/relocate adjacent **entity** (repeated) + determine *derived property* of **entity**.

Task	*Accessibility (entities, constrained)*
Description	Find the set of entities accessible from an entity where the distance is less than n.
Example	Find the *first names* and *last names* of **persons** with no more than three degrees of separation from a **person** with the *first name* "Adam" and *last name* "Smith".
Decomposition	Identify/relocate **entity** with *property* + identify/relocate adjacent **entity** (repeated at most n times) + identify/relocate *property* of **entity** (optional).

Task	*Accessibility (properties, constrained)*
Description	Find a derived property of entities accessible from an entity where the distance is less than n.
Example	Find the *number* of **persons** with no more than three degrees of separation from a **person** with the *first name* "Adam" and *last name* "Smith".
Decomposition	Identify/relocate **entity** with *property* + identify/relocate adjacent **entity** (repeated at most n times) + determine *derived property* of **entity**.

Common connection tasks combine analytical tasks (identify, determine, and relocate) to identify entities that share connections with two or more other entities. In most situations, once connected entities have been found, the user will proceed to study a property of those entities.

Task	*Common connection*
Description	Given a set of entities, find a set of entities that are connected to all of them.
Examples	Find the *first names* of **persons** that have direct or indirect **relationships** of *type* "managed by" to a **person** with the *first name* "Adam" and a **person** with the *first name* "Barbara".
	Find the *types* of direct or indirect **relationships** between a **person** with the *first name* "Adam" and a **person** with the *first name* "Barbara".
Decomposition	Identify/relocate **entity** with *property* (repeated) + identify/relocate adjacent **entity** (repeated) + determine *intersection* + identify/relocate *property* of **entity** (optional).

Connectivity tasks combine analytical tasks (identify, determine, and relocate) to infer knowledge about the connectivity of sub-networks. If *N'* is a sub-network of a network *N*, then every node and every link in *N'* is also in *N*.

Task	*Connectivity (shortest path)*
Description	Determine if two nodes are connected and find the shortest path between them.
Example	Are the **persons** with *first name* "Adam" and *first name* "Barbara" connected?
	Find the smallest *degree of separation* between a **person** with *first name* "Adam" and a **person** with *first name* "Barbara".
Decomposition	Identify/relocate **entity** with *property* (repeated) + identify/relocate adjacent **entity** (repeated) + determine *derived property*.

Task	*Connectivity (clusters)*
Description	Find clusters.
Example	Identify and find the *number* of **cliques** in a social network.
Decomposition	Identify/relocate **derived entity** (repeated) + determine *derived property*.

Task	*Connectivity (connected components)*
Description	Find connected components.
Example	Identify the *number* of disconnected **sub-networks** in a social network.
Decomposition	Identify/relocate **derived entity** (repeated) + determine *derived property*.

Task	Connectivity (bridges)
Description	Find bridges/articulation points.
Example	Find the *first name* and *last name* of the **person** whose removal will result in a disconnected sub-network.
Decomposition	Identify/relocate **derived entity** (repeated) + identify/relocate **entity** + identify/relocate *property* of **entity**.

Attribute-Based Tasks

Nodes tasks combine analytical tasks (identify, determine, and relocate) to infer knowledge about nodes and their attributes.

Task	Nodes (properties)
Description	Find the nodes with specific attribute values.
Example	Find all **persons** with an *occupation* of "manager" and *age* greater than "30".
Decomposition	Identify/relocate **entity** with *property* (repeated).

Task	Nodes (derived property)
Description	Find a derived property of a set of nodes with specific attribute values.
Example	Find the *number* of **persons** with an *occupation* of "manager" and an *age* greater than "30".
Decomposition	Identify/relocate **entity** with *property* (repeated) + determine *derived property*.

Links tasks combine analytical tasks (identify, determine, and relocate) to infer knowledge about links and their attributes.

Task	Links (connected nodes)
Description	Given a node, find the nodes connected by links with specific attribute values.
Example	Find all **persons** with **relationships** of *type* "friend" to a **person** with the *first name* "Adam" and *last name* "Smith". Suppose that links are directional and that they encode managerial relationships; find all **persons** who are *managed by* a **person** with *first name* "Adam" and *last name* "Smith".
Decomposition	Identify/relocate **entity** with *property* + identify/relocate adjacent **entity** with *property* (repeated).

Task	Links (extreme values)
Description	Find the node that is connected by a link with the minimum/maximum value for a link attribute of interest.
Example	Suppose links encode strength of friendship; find the **person** with the *strongest* friendship **relationship** with a **person** with the *first name* "Adam" and *last name* "Smith".
Decomposition	Identify/relocate **entity** with *property* + identify/relocate adjacent **entity** (repeated) + determine *derived property*.

Browsing Tasks

Follow path tasks combine analytical tasks (identify and relocate) to infer knowledge about paths in multivariate networks.

Task	Follow path
Description	Follow a given path.
Example	Find the **person** with *first name* "Barbara" with a **relationship** of *type* "friendship" to a **person** with the *first name* of "Adam"; now find the **person** with the *first name* "Charles" with a **relationship** of *type* "friendship" to her.
Decomposition	Identify/relocate **entity** with *property* + identify/relocate adjacent **entity** with *property* (repeated).

Revisit tasks primarily employ the analytical task relocate to revisit previously visited entities. Typically this is followed with any of the other analytical tasks to infer more knowledge. Although essentially a low-level task, we include revisit here because it is part of Lee et al.'s framework and because we want to emphasize its importance in facilitating explorative analysis [5].

Task	Revisit (entity)
Description	Revisit an entity and infer further knowledge.
Example	After completing the previous task (follow path), go back and find the **person** with the *first name* of "Barbara" and find her other friends.
Decomposition	Relocate **entity** + identify/determine/relocate/compare.

Estimation Tasks

Lee et al. propose a single "overview" task to allow estimation of general network characteristics [5]. This includes estimating the size of the network, the distribution of property values over entities, or getting a rough idea of the clusters in the network. They do not further sub-divide this category. While they state that this is a "compound exploratory task to get estimated values quickly" it is not clear how this task could be precisely decomposed into several component low-level tasks, as doing so would suggest that an exact value for the desired network characteristic could be determined (rather than an estimate, which may, of course, be inaccurate). It also suggests the use of external support in the form of memory, algorithms or computational processing time, since tasks that derive accurate characteristics of entire networks through the use of component low-level tasks can only do so if such external support is used.

Our taxonomy therefore includes "estimation" tasks. We use the term "estimation" (rather than "overview") to emphasize that these tasks are not easily definable in terms of lower-level tasks (as per the Lee et al. definition [5]), but are high-level, with the objective of gaining a rough estimation rather than precise answers. In this sense, there is also a clear link with the "cognitive" ("infer") task category of Valiati et al. [11], although Valiati et al., like Lee et al., suggest that these tasks can be systematically decomposed into sequences of sub-tasks.

The definition of our estimation tasks is based on the premise that external support is not available during task execution, and that precise answers are therefore not possible. The alternative would be to define overview/inference tasks algorithmically in terms of the use of low-level tasks, memory storage and computations so as to ensure accuracy; this systematic approach would add little to what is already known about task decomposition. Since neither Lee et al. nor Valiati et al. have provided sub-categories for their overview/inference tasks, we introduce categories for estimation tasks below. The sample of general information visualization tasks defined by Amar and Stasko describe the types of high-level objectives a user may have [3] (also see the introduction), and these are used in distinguishing two types of estimation tasks: *understanding* and *comparison*.

Understanding task have the aim of gaining more complete understanding of the information; they relate to the Amar and Stasko tasks of decision making, learning and identifying domain parameters [3].

Task	*Clusters*
Description	Characterize sets of nodes as (potentially) belonging to highly-connected groups (clusters).
Example	In a social network, identify all those people who are likely to attend parties held by Adam, Barbara, and Charles. This task requires identifying a cluster of nodes for each of Adam, Barbara, and Charles. These clusters may overlap, and some nodes in the network may not belong to any of these three clusters.
Explanation	This task identifies groups of nodes that are structurally highly connected; no use is made of attribute information. The estimation is based on scanning the network structure, identifying sets of nodes that are closely linked. The definitions of the clusters may be inaccurate unless the entire network is systematically and algorithmically analyzed to identify which sets of nodes form tight clusters, while keeping a record of all the connections. An estimated cluster may therefore include nodes that are only related to some (but not many) members of the cluster; or may omit some nodes that ought to be members.

Task	*Common attributes (nodes)*
Description	Characterize sets of nodes as belonging to different groups, based on node attributes.
Example	In a social network, identify all the girls who live in Glasgow, who have blue eyes, who are over 17, and who play tennis. This task is concerned with the values of five different attributes; the result is the set of nodes for which these values match the specification.
Explanation	This task identifies groups of nodes that share similar characteristics, based on several given attribute/value pairs; no use is made of structural information. The estimation is based on scanning the nodes and their attributes, identifying groups of nodes with the same attribute values. The definitions of the groups may be inaccurate unless all the nodes are systematically and algorithmically inspected to determine the values of their attributes, keeping a record of the nodes and their values. An estimated group may therefore include nodes that have only some of the correct attribute/value pairs, or may omit nodes with all the specified characteristics.

Task	*Common attributes (links)*
Description	Characterize sets of nodes as belonging to different groups, based on link attributes.
Example	In a network representing people and the email communications sent between them over the course of a week, identify all the people who sent humorous emails on Monday morning. This task is concerned with the values of the attributes associated with the links: the email content and its date.
Explanation	This task identifies groups of nodes that share similar relationships to any other nodes, based on given attribute/value pairs of their associated links. The estimation is based on scanning the nodes and their relationships (and the attributes associated with their relationships), identifying those entities associated with the correct type of relationship. The definitions of the groups may be inaccurate unless all the links are systematically and algorithmically inspected to determine the values of their attributes, keeping a record of the associated nodes. An estimated group may therefore include nodes that are not associated with the correct type of relationships, or may omit nodes that do.

Task	*Domain (nodes)*
Description	Determine the attributes and values associated with nodes.
Example	Identify all the attributes used for nodes, and all their possible values.
Explanation	This may be inaccurate unless all nodes are visited systematically or algorithmically to extract and record their attributes and values.

Task	*Domain (links)*
Description	Determine the attributes and values associated with links.
Example	Identify all the attributes used for links, and all their possible values.
Explanation	This may be inaccurate unless all links are visited systematically or algorithmically to extract and record their attributes and values.

Comparison tasks are concerned with understanding changes in a network, and relate to the Amar and Stasko tasks of identifying trends and causation, prediction, hypothesis verification, discovering correlative models, and seeing the effect of uncertainty [3]. These tasks assume the existence of more than one instance of a network, each representing a different point in time. For completeness, we include comparison tasks here, but a more detailed discussion of temporal networks is deferred to Chap. 8.

Task	*Trends*
Description	Compare information at different stages in a changing network.
Example	In a social network, characterize how the group of friends centered around Adam changes over the course of a year.
Explanation	A changing network is described as a series of time-slices, where each time-slice is an instance of the network.
	This result of this task is a description of how the network has changed between two (or more) of its time-slices. Typically, it would be overview information (as described in the five "understanding" tasks above) that is compared, rather than specific node/link information.
	This comparison will result in uncertain information unless external algorithms are used to explicitly compare the details of the information in the series of networks.

Task	*Causation*
Description	Formulate an explanation why two time-slices in a changing network are different.
Example	Explain why some students were friends with John (the smartest student in the class) the week before an assignment was due, but not the week after.
Explanation	This task is different from the others listed above, as it requires external knowledge, that is, information that is not represented directly in the network itself.

5.3 Discussion

The approach that we have taken in this chapter is to review the relevant literature to come up with a pragmatic synthesis of other frameworks, with particular reference to multivariate networks. In doing so, we have considered ideas from general information visualization methods [12], general information visualization tasks [3], specific information visualization questions [2], multi-dimensional visualizations [11], and visualization tasks for univariate graphs [5].

It is worth reflecting on how the framework presented here corresponds to other recent work on information visualization tasks. As noted before, Schulz et al. describe the visualization task design space along five dimensions (goal, means, characteristics, target, and cardinality) [7], while Brehmer and Munzner consider three questions (why?, how?, and what?) [4]. Although these frameworks are much more general than multivariate networks and somewhat abstract for a general audience, they provide a very useful approach to reflect on some of the key points discussed in this chapter.

Goal, or *why?*, corresponds with the notion of user intent. Users study multivariate networks to gain insight about the phenomena, such as social networks, that they describe. Brehmer and Munzner emphasize that, depending on the context, there will be different levels of specificity of tasks. For example, they distinguish between high-level (consume), intermediary-level (search), and low-level (query) objectives. Schulz et al. point out that visualization supports exploration, confirmation of hypotheses, and presentation of findings. In the light of supporting interactive analysis, our emphasis has been on the former two (exploration and confirmation). However, we note that other objectives such as presentation, communication, or even interaction with a visual representation of a data set as a form of entertainment are all valid.

Means, or *how?*, describes how a task is carried out and relates to the operational and analytical tasks described in this chapter. Most of our attention has gone into describing these tasks for a multivariate network context.

Characteristics, target, and *cardinality,* or *what?*, are concerned with how the data relates to the task. The notion of *characteristics* distinguishes between low-level and high-level aspects of the data. This corresponds closely to the difference between tasks where knowledge is directly derived from the data (for example, structure- and attribute-based tasks) and ones that require more nuanced deduction and uncertainty, as highlighted by our estimation tasks. *Target* highlights the parts of the data on which analysis focuses. This chapter picks up on this by emphasizing the entities of multivariate networks (nodes, links, paths, and networks) as well as associated properties (structural properties and attributes) and how tasks relate to these. In the context of this chapter, *cardinality* emphasizes that tasks may include the analysis of single or multiple networks. Although Brehmer and Munzner leave the question of *what* rather open-ended, they do emphasize the importance of defining the inputs and outputs associated with a task, especially when several tasks are combined sequentially. We do not treat this issue explicitly, but our examples imply that for our purposes the inputs are multivariate graphs and the outputs are subsets of entities (nodes, links, paths, and networks) and/or properties (structural and attributes).

Finally, it should be noted that all considered tasks necessarily involve a visual representation of one or more multivariate networks, and interaction with this visual representation. This chapter has not tried to describe interaction methods, such as filtering and zooming, which have been bundled under the configure task. We note, however, that the distinction between task and interaction method is not always clear-cut and many authors have chosen to combine examples of both (for example, [8]). Chapter 6 provides a more in-depth analysis of interaction methods for multivariate graphs.

5.4 Conclusion

In this chapter we have described tasks for multivariate networks. We have summarized the entities and properties of multivariate networks and presented a general taxonomy for visualization tasks. We then described a task framework specifically for multivariate networks and showed how the proposed tasks can be composed of lower-level tasks of the general taxonomy. We also discussed some of the implications of this framework in the light of related work on information visualization tasks.

Many of these tasks (in particular the estimation tasks) have been defined without consideration of any context or users' prior knowledge. In future work, a more semantic and situational analysis of tasks relating to multivariate networks might take into account how such knowledge might affect the way in which tasks are executed and their results interpreted. Examples of such contextual knowledge could include related node attributes (redheads tend to have blue eyes), assumed edge attributes (people tend to like their children), or broader population attributes (most computing science graduates are male).

Almost invariably, research on visualization tasks is motivated in two ways. First, an understanding of a domain problem should be translated into user tasks to support. The user tasks, in turn, should have a direct bearing on the design of a visualization system to address the original domain problem. Second, an understanding of user tasks enable visualization designers to evaluate the suitability of their designs and systems in addressing a domain problem. Our aim in this chapter has been to provide an introduction and overview of tasks for multivariate network analysis to a general audience and, hence, we have not evaluated the suitability of our framework to support these objectives. We suspect that it may be useful to this end, but further work is required to make this claim.

Acknowledgements

The authors wish to thank Peter Eades (University of Sydney, Australia), Helen Gibson (Northumbria University, United Kingdom), Daniel Keim (University of Konstanz, Germany), and Robert Kosara (Tableau Software, United States) for fruitful discussions and constrictive input at the Dagstuhl Seminar *Information Visualization – Towards Multivariate Network Visualization* held at Schloss Dagstuhl, Germany, from 12 May–17 May, 2013. A.J. Pretorius was supported by a Leverhulme Early Career Fellowship under award number ECF2012-071.

References

1. Ahn, J., Plaisant, C., Shneiderman, B.: A task taxonomy of network evolution analysis. Tech. Rep. HCIL-2012-13, Human-Computer Interaction Lab, University of Maryland (2012)
2. Amar, R.A., Eagan, J., Stasko, J.T.: Low level components of analytic activity in information visualization. In: Proceedings of the IEEE Conference on Information Visualization, pp. 111–117 (2005)
3. Amar, R.A., Stasko, J.T.: Knowledge precepts for design and evaluation of information visualizations. IEEE Transactions on Visualization and Computer Graphics 11(4), 432–442 (2005)
4. Brehmer, M., Munzner, T.: A multi-level typology of abstract visualization tasks. IEEE Transactions on Visualization and Computer Graphics 19(12), 2376–2385 (2013)
5. Lee, B., Plaisant, C., Sims Parr, C., Fekete, J.D., Henry, N.: Task taxonomy for graph visualization. In: Proceedings of the AVI Workshop on Beyond Time and Errors: Novel Evaluation Methods for Information Visualization, pp. 1–5 (2006)
6. North, C.: Toward measuring visualization insight. IEEE Computer Graphics and Applications 26(4), 6–9 (2006)
7. Schulz, H.J., Nocke, T., Heitzler, M., Schumann, H.: A design space of visualization tasks. IEEE Transactions on Visualization and Computer Graphics 19(12), 2366–2375 (2013)
8. Shneiderman, B.: The eyes have it: a task by data type taxonomy for information visualizations. In: Proceedings of IEEE Visual Languages, pp. 336–343 (1996)
9. Shneiderman, B., Aris, A.: Network visualization by semantic substrates. IEEE Transactions on Visualization and Computer Graphics 12(5), 733–740 (2006)
10. Spence, R.: Information Visualization: Design for Interaction, 2nd edn. Prentice Hall (2007)
11. Valiati, E.R.A., Pimenta, M.S., Freitas, C.M.D.S.: A taxonomy of tasks for guiding the evaluation of multidimensional visualizations. In: Proceedings of the 2006 AVI Workshop on Beyond Time and Errors: Novel Evaluation Methods for Information Visualization, pp. 1–6 (2006)
12. Wehrend, S., Lewis, C.: A problem-oriented classification of visualization techniques. In: Proceedings of the IEEE Conference on Visualization, pp. 139–143 (1990)
13. Yi, J.S., Kang, Y., Stasko, J.T., Jacko, J.A.: Toward a deeper understanding of the role of interaction in information visualization. IEEE Transactions on Visualization and Computer Graphics 13(6), 1224–1231 (2007)

6

Interaction in the Visualization of Multivariate Networks

Michael Wybrow, Niklas Elmqvist, Jean-Daniel Fekete, Tatiana von Landesberger, Jarke J. van Wijk, and Björn Zimmer

The overall aim of visualization is to obtain insight into large amounts of data. Detection of patterns as well as outliers are typical examples. For networks, such patterns can be number and position of cliques; for multivariate data this can be the correlation between attributes. The major challenge of multivariate network visualization is to understand the interplay between properties of the network and its associated data, for instance to see if the formation of cliques can be understood from attributes of nodes.

Producing useful and informative visualizations for multivariate networks is a complex and challenging task. Complexity and scalability (see Chap. 10) are significant issues, both with respect to the graph size as well as to the number and variety of variables. It is very difficult to statically display large, complicated data sets in general, including multivariate data and networks. Occasionally it is possible to nicely encode small multivariate data sets completely in custom static visualizations, such as with Minard's seminal "Napoleon's March to Moscow" visualization [53], but this is rare.

In practice, even moderate-sized networks can be difficult to visualize without overlaps and loss of information, let alone when augmented with additional variables. Moreover, people working with visualizations can usually only comprehend a small subset of the information space at a time. It is therefore important to reduce the relevant information displayed at any point to a manageable amount in order to facilitate understanding of the main data characteristics. Thus, as the data size and complexity (i.e., the combination of dimensions and network complexity) increases, there is a need to efficiently navigate through the data and to enable discovery and communication of the data.

Interaction is a vital component in the visualization of multivariate networks. By allowing people to browse data sets with interactions like panning and zooming, we can enable much more information to be seen and explored than would otherwise be possible with static visualization. Overview-based interactions afford the user the ability to understand a complete picture of the data or information landscape and to decide where to direct her attention. Through search and filtering, interaction can reduce cognitive effort on users by allowing them to locate, focus on and understand subsets of the data in isolation. Pivoting and other navigational interactions at both the view and data level allow people to identify and then to transition between areas of interest.

While there are methods for interacting with graphs and dimensions separately, the combination of both needs special attention. The challenge is to clearly visualize multiple sets of individual dimensions as well as to offer a useful visual overview of data, and allow transitions between these to be easily understood. Moreover, we need to find ways to support users in navigating through the complex data space (graphs × dimensions) without "getting lost," and without an overburden of interaction actions that may frustrate the user.

In this chapter interaction for the visualization of multivariate networks is considered. After a discussion of the design space for interaction, existing approaches are examined, guidance for designing interactions is offered and open problems in the area are described. It is aimed at readers who are intending to visualize networks with multivariate data. They may be planning to evaluate and select some existing approaches or systems and adapt these to their needs, or they may be thinking about designing a custom visualization tailored to the needs of their data and audience. Rather than just a survey of the field, this chapter should be considered a guide to interaction for networks with multivariate data; explaining what the problems are, what is possible, what has been done before, what might be done in future.

The rest of the chapter is organized in five further sections. The next section discusses the design space and requirements for working with large multivariate data sets, including difficulties in navigating networks and dimensions. Section 6.2 classifies relevant interaction techniques on the basis of the stages in the standard Information Visualization Reference Model. Section 6.3 gives examples of the interactive aspects of multivariate graph visualization systems. Section 6.4 presents recommendations and guidelines for designing novel interaction approaches, including adaptation of existing interaction design principles for use in this setting. Finally, Section 6.5 puts forward a vision of the challenges and goals as we see them within the field of multivariate graph visualization.

6.1 Background

Interaction is a vital ingredient of information visualization, and has been heavily studied. In this section, we do not aim to explain in general how interaction works in visualization, as this is very well addressed by excellent books such as [58] and a large number of articles [30, 39, 72]. Also, we acknowledge that data exploration encompasses much more than just direct interaction with graphical representations, and includes aspects like navigation support, knowledge capture, and collaborative visualization. This area is studied in visual analytics; for an overview see Pike et al. [51].

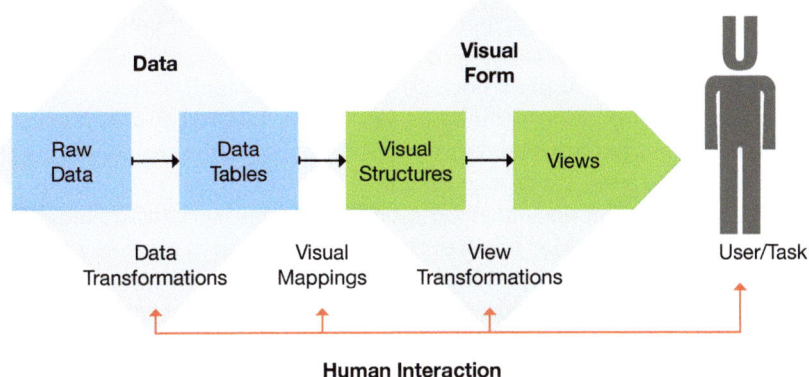

Fig. 6.1. The information Visualization Reference Model

Furthermore, for this chapter, we mostly consider interaction for standard point-and-click and keyboard interfaces on desktop computers. While multi-touch tablets are commonplace and we are seeing increasing availability of large touch-based tables and displays, there has been relatively little work designing or evaluating interaction techniques for working with large networks or multivariate data on these. This is also the case with other new technology now becoming available to consumers such as 3D displays, contactless input devices, and multi-monitor displays. We discuss this as a key ongoing challenge in Sect. 6.5.

Data exploration often involves a top down approach, as strongly summarized in the visual information seeking mantra of Ben Shneiderman [56]: "overview first, zoom and filter, details on demand". Both for network and multivariate visualization, many systems and techniques aim to satisfy this pattern. But in practice, a bottom up approach is used. For instance, in social network visualization a certain person can be the starting point for further exploration [27]; in multivariate visualization one can start from one particular item and explore items which are similar. Since these approaches are valuable, an ideal system should support both.

To describe the multiple kinds of interactions used for the visualization of multivariate networks in more detail, we use the Information Visualization Reference Model [10] (see Fig. 6.1), which breaks down the visualization process into four stages: *raw data* or *source data*, *data tables*, *visual structures* or *visual abstractions*, and *views*. To display the raw data, several transformations have to be applied: the raw data is transformed into data tables through data transformations, the data tables into visual structures through visual mappings, and the final rendering transforms the visual structure into an image on a view. All these transformations are performed using a multitude of specific parameters, and interaction can then be defined at the system level as *the change of transformation parameters controlled by the user with immediate feedback to the user.*

This generic model applies both to network and multivariate visualization, and many interaction techniques specifically tailored to the properties of these data types have been developed. In the next section we enumerate the most relevant of these, categorized along the stages of the reference model. However, far less techniques have been developed that specifically aim at interaction with combinations of network and multivariate data. The challenge here is to offer a simple but powerful set of interaction techniques that allows users to explore such combinations with minimal cognitive overload. On the one hand, this should be achievable, since many tasks and operations are similar at a high level; but on the other hand, standard representations of networks and multivariate data do vary largely, and also the more powerful and customized interaction methods for dealing with these data types differ greatly.

These effects can be observed for all stages of the reference model. At first sight, network data and multivariate data seem fundamentally different. However, topological aspects of network data can be nicely captured as multivariate data, simply by calculating topological metrics of nodes and edges. Also, multivariate data can be considered as networks, for instance by introducing edges between nodes that are similar, as pursued by Liu et al. [42]. Having said this, multivariate network visualization usually cannot be reduced to purely multivariate or network visualization. In fact, the combination makes analysis of multivariate networks a real challenge since discovery of an underlying phenomenon in the data can require a detailed understanding of the network topology together with the multivariate attributes, e.g., if variables represent snapshots of a flow dictated by the topology. One consequence for interaction is that users should be enabled to obtain such associated data on request. Filtering of data is a standard operation. For multivariate data this typically involves selection based on ranges of attributes, and for networks the distance from a selected set of nodes can be used.

Concerning the visual representation, network data and multivariate data can be shown separately or be combined. The use of multiple views on data is standard in visualization, and by interacting through linking and brushing, information from different views can be associated. Interaction is crucial here, but also, as both types of data are shown separately, fusion of information is often hard. One way to provide a combined view is to use a network-based approach, where nodes and edges are embellished with iconic representations of values or attributes. This limits the use of standard interaction methods for multivariate data, for instance, to select two ranges for attributes by sweeping out a rectangle in a scatterplot. Another way to combine data in one view is to use a multivariate data-based approach, for instance, by superimposing edges on top of a scatterplot. Now, standard interaction methods for multivariate data can be used, as positions of nodes encode attribute values, but also, some network interaction techniques that imply changes in the layout cannot be used anymore.

The standard approach in the view transformation stage is to provide options for zooming and panning. On the image level, this is straightforward, however, when using multiple views where the spatial dimensions have different meanings, this can be hard to deal with in a natural way.

These examples show there are basically two approaches to interacting with multivariate networks. One approach is to stick to conventional representations and dedicated interaction methods, another, more challenging but also potentially more rewarding approach is to aim for tight integration, both with respect to representation and interaction, to facilitate the understanding of the relation between network and multivariate. In the following sections these approaches are explored in more detail.

6.2 Classification of Interactions

We use the Information Visualization Reference Model, originally presented in [10], as the basis for our classification of interaction techniques (see Fig. 6.1). We classify interaction techniques based on the level of this pipeline they affect. Note, the match may not be always perfect, as some techniques address multiple levels simultaneously. Where possible, we make use of standard terminology and jargon from the information visualization community in order to simplify access to related work.

Notably, our classification presents the pipeline stages in the reverse order to [10]: we describe interactions at the view-level first for pedagogical clarity, since these are simpler, easier to understand, and are sometimes extended or utilized by interactions in the remaining stages of the pipeline.

Many of the generic interaction techniques are applicable both to standard networks as well as multivariate data, and basic examples are given. As discussed in Sect. 6.1, there are many possible graph representations, the choice of which can limit the applicability of interaction techniques since these may be dependent on specific aspects of the chosen graph representation. Examples of complete systems utilizing a mix of interaction techniques to deal simultaneously with a combination of multivariate data and networks are described in Sect. 6.3.

This classification is a revised and expanded version of a similar classification of interaction techniques for network visualization appearing in [68]. Note, this is certainly not the only way to define and categorize interaction. For instance, Yi et al. advocate for a taxonomy based on user intent, and they distinguish Select, Explore, Reconfigure, Encode, Abstract/Elaborate, Filter, and Connect as main categories [72]. Similar classification has been recently also presented for cartography [54].

6.2.1 View-Level Interactions

The view-level interactions are mostly related to visual emphasis of interesting objects, navigation through the data set, and using Magic Lenses to augment the visualized information.

Highlighting

Highlighting transiently changes the visible rendition of items at the view-level, not at the visual encoding level. Although it can be practically implemented with support at the visual structures level, this is not required so we conceptually consider it a view level interaction.

Interactions such as search or mouse hovering may lead to highlighting of objects such as search results or linked content.

Hovering

Hovering is used in multivariate visualizations such as InfoZoom [61] that display large data tables with a smart aggregation mechanism. Rows are items, columns are attributes, and values are in cells. When the mouse passes over a value in a cell, all the cells with the same value for that attribute are highlighted, showing the frequency and distribution of this value. Hovering is even more useful with multiple views to highlight parts linked by some relation. MatrixExplorer [31] uses two linked visual representations for networks, one being a node-link diagram and the other an adjacency matrix. When the mouse hovers over an entity in one visual representation, the same entity is highlighted in the other.

Brushing and linking

This technique involves the user watching multiple views related to the same dataset. When the pointer is moved over an item in one view, all the related items are highlighted in all the views [4, 9].

For multivariate networks, these views can use the same visual representation or a different one; they can show the same information (e.g., the network topology as a node-link diagram as well as an adjacency matrix [31]), complementary information (e.g., the network topology as a node-link diagram and nominal attributes as lists [28]), or mixed aspects (e.g., the network topology as a node-link diagram and attributes using parallel coordinates [3, 60]). These can be used to more easily contrast and compare information or variables in distant places within the network or to see parts of the dataset from different perspectives. The latter is often used when visual encoding does not allow for viewing all the data in one visual representation. This is generally caused by data size (too many data points to show) or data complexity (too many data variables).

Further interaction techniques are often used to augment and enhance the use of multiple views. Some of these will be described in the visual encodings section.

6.2 Classification of Interactions

Magic lenses

Magic lenses [7] are "filters, that modify the presentation of application objects to reveal hidden information, to enhance data of interest, or to suppress distracting information." They have been used extensively in visualization of networks and multivariate data.

Excentric Labeling [20] offers an approach similar to tooltips: labels are interactively displayed over dense visualizations such as scatterplots or node-link diagrams. When enabled, they show a focal region (rectangular or circular) that follows the mouse; all the items inside the region are labeled outside of the region with a line connecting each item to its label. Bertini et al. [5] has extended upon this to give better control of the focal region and visualization of aggregated information on the focal region.

Jusufi et al. [36] describe lenses for multivariate network that display nodes as small multidimensional visualizations when they are within the focal area. They use several visualizations: parallel coordinates, bar charts, and star plots.

Navigation

Panning and zooming

Panning and zooming involve changing the visible viewport over the otherwise unchanged visualized data. These actions are usually accomplished via standard interactions with common controls like scroll bars and sliders, hardware like mouse scroll-wheels and track-pads or using multi-touch at touch tables or tables.

Several navigation techniques have been designed to improve panning and zooming over large data sets, which are discussed in detail in Chapter 10. Suffice it to say that these operations can be very cumbersome, requiring users to drag the cursor for long distances across the screen. The simplest technique to overcome that problem is to use an overview plus details representation, such as a bird's eye view of the visualization in a small window and detailed view in a large one. The viewport of the detailed view is usually displayed as a rectangle on the small window that can be manipulated for fast panning.

In graph visualization, *topology-aware graph navigation* allows automatic panning and zooming in a graph. These actions can be performed directly on the network structure, such as link sliding [46] or bring-and-go [63]. These techniques allow the user to quickly find out-of-viewport nodes that are attached to a particular node, relocate these to be temporarily positioned in their current view and then allow further navigation from them. The bring-and-go technique can also be considered as a magic lens for navigation.

View distortion (single/multiple)

View distortion allocates more space to items of the users' interest. In particular, fisheye views generally allow people to see more information at a point of interest. For example, this can reveal detailed information that was initially smaller than one pixel in size.

For graphs, there are specific distortion techniques, such as Balloon Focus in a treemap [64] and a guaranteed visibility technique in dendrograms [47] that allocate more space to the nodes in focus for their detailed inspection. These techniques allow for multiple foci at the same time.

Distortion can be applied also to edges, improving the visibility of items on the screen. For example, Edge Lenses [71] interactively displace edges under the pointer in order to avoid overplotting of edges over nodes or edges over each other. Tominski et al. [63] have proposed two types of lenses to facilitate the exploration of networks: *Local Edge Lens* only show edges with vertices inside the focal region to locally reduce clutter; *Bring Neighbors Lens* transiently moves vertices that are connected to vertices in the focal area but not visible in the viewport at the boundaries of the focal area. Their lenses can also be combined. Note that the latter technique can be seen as an example of magic lenses.

The view distortion is not always geometric: *Semantic zooming* changes the visual representation and level or details according to the zoom level. The interaction technique remains the same as panning & zooming (e.g., using the mouse wheel or a zoom slider) but the visual effect of zooming is changed. Semantic zooming [50] involves changing the visual parameters by altering the amount of detail shown at various levels of zoom. The simpler kind of semantic zooming consists of showing more details when zooming in, and less when zooming out, connecting the zoom level to the data aggregation level [19]. This could involve showing more of a network at the greater zoom depth such as changing graph aggregation level [17].

6.2.2 Visual Structure-Level Interactions

Selection

Selection interactions alter the visual parameters of the visualization. They generally result in the most basic form of encoding change in order to highlight or emphasize areas of the network. Often they modify visual attributes of the graph entities (e.g., color, size, line width, etc). Selection differs from the view-level highlighting in that it implies a state change at least at the visual structure level, sometimes even at the data level. Also, highlighting is transient and changes implicitly as the pointer moves or the search query changes, whereas selection is explicitly set on or off.

There are various ways of selecting. For graphs one can select/brush nodes directly by clicking on them, select nodes according to their network properties [6] or select items according to network attribute values. The latter is specifically suitable when analyzing multivariate networks. Moreover, the network structure can be used for an enhanced highlighting, i.e., not only the selected nodes are highlighted but also their neighbors or parent/child nodes. This can be extended with node or edge properties, where only those adjacent/connected nodes are highlighted that have certain node attribute values. An example is highlighting of controlled companies in a shareholding network [62].

McGuffin et al. [44] have described techniques to select subgraphs interactively. In addition to traditional rectangle and lasso selection of nodes, they introduce a special kind of radial menu to further control and extend the selection of nodes (e.g., extending it by increasing a radius from the current selection: add nodes at distance 2, 3, etc.) They also introduce a special kind of menu box that appears transiently to operate on the current selection for visual structure level or data level operations (e.g., align, change color, change shape, etc.).

Changing mapping of attributes

Interactions that change the visual encoding can also be used to explore and understand various dimensions of the data. An example of this is changing the visual mapping of attributes, i.e., which attributes are assigned to which visual attributes such as size and color. Such interaction should be typically provided in interactive graph visualization systems.

Even considering just classic node-link representations for networks, visual encodings and styles of these may still vary greatly. Different emphasis can be given to visual objects, such as by drawing edges faintly using a high level of transparency or displaying nodes as points without size. These choices can in turn lead to vastly different visual results for the same data. Hence, interactively varying such attributes of the visual encoding can be useful to discover different properties of the data. See [41] for some of the more extreme examples, as well as further discussion of similar techniques in Sect. 10.2.2.

Network layout

Layout-based interactions alter the position of nodes and edges based on properties of the network. The intent is for the layout to reveal additional information about the structure of the network.

Examples of layout-altering interactions include positioning nodes and edges to emphasize similarity, such as using Multidimensional Scaling [40], or by applying existing automated graph layout algorithms. Interactions to apply layout changes are typically triggered by changing a layout setting, however layout can sometimes be adjusted by interacting directly with the network, i.e., dragging nodes or edges.

Network layout can be calculated solely in dependence of network structure [26], only in dependence on node properties [6] or a combination of both network structure and network attributes [37, 57]. The type of layout depends on the user task. If the user wishes to analyze the relationships between nodes in the network, a topology-only layout is sufficient. However, if she wishes to analyze the interconnection of network structure and network attributes (e.g., are people with similar characteristics friends?), a layout that takes both network structure and network attributes into account is preferable.

Moreover, constraint-based network layout approaches can allow interactive control and fine-tuning of the layout [16], and may be used in conjunction with multiple views and semantic zooming to allow interactive browsing and exploration of large multivariate networks [15].

Multiple differing network layouts can be coupled with multiple views and augmented with brushing and other highlighting techniques to understand the relations between them [11]. This allows the user to compare and analyze the network from different perspectives, and detect information which might have been hidden while using a single layout.

Representation

Graphs and multivariate data can be represented visually in various ways (e.g., node-link diagrams vs. adjacency matrices for graphs; scatterplot matrices vs. parallel coordinate plots for multivariate data, etc.). As one representation may not reveal the intended information on the network, the user may wish to change the representation in order to gain a better view of the data. This is done using interactions altering visual encoding of parts of the network or present alternative representations such as matrix views, tables, or even a mixed representation such as in NodeTrix [32].

6.2.3 Data-Level Interactions

Data-based interactions involve selecting which data to show (showing more, less or completely different data) or manipulating data values (deleting, inserting data).

Selecting Data for Visualization

Filtering

For large graphs, the whole graph may not be shown on the screen. The user then can decide either to reduce the size of the displayed data set (filtering) or to expand on demand the currently shown part of the data set (adding undisplayed data). Then, data level filtering interaction enables display of just interesting subsets of the data.

Such interaction can be performed directly in the network visualization (by selecting nodes to hide) or using a query interface. The query interface can range from a simple slider for attribute values, to a histogram-based filter, right through to filtering via brushing in additional views on the data (multiple views).

Dynamic querying

Sometimes there can be one or more important variables to focus on within the visualization. A prominent example is time. The user may wish to browse through time in the visualization of dynamic graphs. For this it is useful to provide controls allowing the user to directly move through the range of possible values. This is analogous to using sliders and other common controls to provide panning and zooming for the space dimension.

Adding undisplayed data

An alternative way of exploring large graphs is to show a small part of the graph at the beginning of the analysis process (e.g., as a result of a search for interesting nodes) and then expand this selection on demand [27]. The expansion allows the user to add undisplayed data to the network. This can be by navigating through the network topology, such as showing neighboring nodes or connections between nodes on demand [28]. In hierarchic graphs (trees), one can navigate along the hierarchy and show nodes on a lower level of hierarchy, or show only nodes at a certain level [19].

The number of possible expansions of a graph might be very large, and the user may not know which parts of the graph to expand. In such situations, it is useful to show information on which elements to display when there are more candidates than there is room to show. Such decisions are often based on a *degree of interest* function. Such functions can be calculated in many different ways (e.g., [23, 27, 29, 43]).

Search

Search-based interactions at the data level are most useful when not all of the multivariate network data can be shown at once. They allow particular entities of interest to be extracted and displayed or highlighted from the entire data set. Specific examples are to:

- search for nodes/edges with certain attribute values;
- search for nodes/edges with certain topological properties;
- search for subgraphs with specific structural properties (motifs) [67]; and
- search for graphs—interactive user interfaces for defining query graphs and searching for them [66].

Search actions may be performed in various ways. They may involve construction of textual or graphic queries, may be performed by example, or achieved by finding similar items to those in a selection drawn or otherwise specified by the user. Search interactions may result in other data level changes such as filtering and adding undisplayed data.

Pivoting

In the case that different variables are represented by different edge and node types in a heterogeneous network, pivoting is an interaction approach where the user can visualize a couple of variables at once and switch between looking at various slices of the data. Usually this involves keeping some common part of the network visible and as stable as possible during pivot actions, such as in PivotGraph [69] or PivotPaths [13].

Changing Data

In some cases, the user may wish to change the input data such as data attributes or graph structure. This can be done by direct data editing in the user interface or by aggregation.

Editing

Multiple different ways to edit graphs exist:

- Graph structure: the user may wish to edit the graph structure: delete or add nodes or edges. This is usually done directly in the visual interface by selecting nodes/edges to delete or by drawing new nodes or edges. The system can then either show these changes directly or can show the impact of these changes on the network structure [67].
- Attributes: in multivariate networks, the user may additionally change attribute values for certain nodes. Moreover, the user can run a specific algorithm which creates new attribute values that can be explored or used for navigation in the graph. This includes creating new attributes by combining existing attributes (such as sum of two attributes) or by creating attributes describing node or edge topological information (e.g., betweenness centrality).

Aggregation

Large graphs are often simplified by aggregation. Aggregation merges several nodes and/or edges to so-called supernodes or superedges, where a supernode represents several nodes and a superedge represents several edges. The user may choose to see one of the predefined graph levels (pre-defined aggregation) or define the aggregation interactively. Such aggregation can merge user-selected nodes into one node [2] or can automatically merge nodes based on user-defined node attributes [69] or on topologic network properties [67]. The aggregation based on selected network attributes is specially useful for multivariate networks. This allows for variable views on the graph and its structure (also cf. Sect. 10.2.1).

Annotation

Annotation is an interaction where the user can add additional information to objects in the visualization in order to augment their understanding of the data and indicate or signpost points of interest. This is analogous to using notes in order to make sense of complexity, although this is arguably more valuable when it is done in-place by annotating the network directly. In this way the annotations cause changes to the data which subsequently allows the user to search, filter or otherwise interact with the annotations directly.

History and provenance

Interactive exploration and analysis of large graphs includes many steps—interaction actions—and feedback loops. The performed interactions are then difficult to remember and reproduce. This is facilitated by tracking of user actions. GraphDice [6] records view changes and selection changes and shows them as a set of miniatures. Hovering over a miniature transiently changes the selection to use the one recorded in the history. Clicking on the miniature sets the view and selection to the recorded one. The RelaNet System tracks and automatically aggregates all user actions [65]. It then shows them to the user using a graphical representation: a tree whose nodes are visualization states and edges are actions. The user can click on a node in the tree in order to resume that previous visualization state. The user can then either replay the actions or start a new exploration path (creating a new branch in the tree). All actions can be stored, shared, and reviewed.

Recorded actions can be analyzed algorithmically or shown to the user for their visual inspection. The CZSaw system [38] keeps track of all interactions and allows the user to explore and share them.

The tracking, reproducibility and analysis of user actions is still a large challenge in visual analytics. This problem belongs to the more general issue of analytical provenance[1] addressed by systems such as VisTrails [59].

6.3 Exemplars

To better illustrate effective interaction techniques and methodologies for multivariate graphs, we here present four exemplars of existing InfoVis systems that include such techniques. These exemplars are the GraphDice system by Bezerianos et al. [6], the GraphTrail system by Dunne et al. [14], Parallel Node-Link Bands by Ghani et al. [22], and the state transition networks by Pretorius and van Wijk [52]. We used the following criteria when selecting these exemplars for inclusion in the chapter:

---

[1] See http://www.vacommunity.org/AnalyticProvenanceWorkshop for the first workshop on this issue.

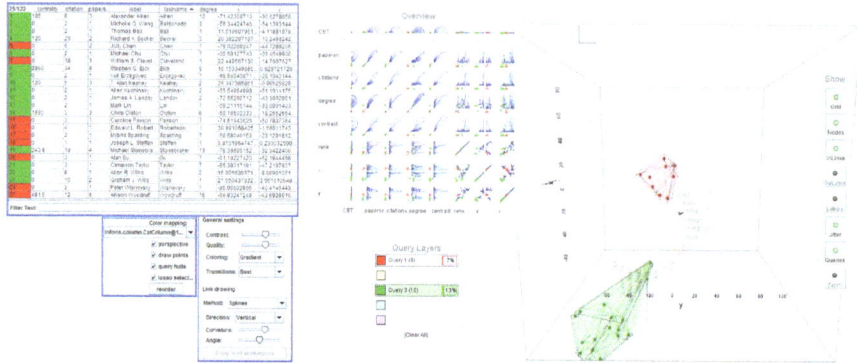

Fig. 6.2. The GraphDice [6] multivariate visualization tool shown visualizing an IEEE InfoVis co-authorship network consisting of both intrinsic and derived attributes. The analyst is in the process of transitioning between two different node attributes; the transition is shown as a smooth animation.

- **Representative:** Our objective was to select exemplars that capture a wide range of representative interaction techniques.
- **Significant:** The included examples all provide interaction techniques that are among the first of their kind.
- **Best practices:** All exemplars demonstrate best practices in interaction for multivariate graphs.
- **Familiar:** Our selection is by necessity limited by our knowledge, experience, and preconceptions of the general field of multivariate graph interactions.

In no way do we claim that this set of exemplars is exhaustive or optimal. There may exist several other InfoVis tools that we could have selected instead of these four. We only claim that our selection is representative and illustrative.

We use the term "analyst" to refer to a domain specialist performing analysis tasks with the system, rather than a "data analyst".

6.3.1 GraphDice

GraphDice [6] is a multivariate graph visualization tool that supports navigation in data space similar to the scatterplot matrix navigation proposed in the ScatterDice [18] tool. The key contribution of GraphDice is the integration of attribute-based layout with interactive data space navigation, where both intrinsic (such as the age, gender, and annual income) as well as derived (layout position, degrees, and centrality) attributes of actors in a social network form the data space. This supports a smooth and fluid visual exploration process where users can seamlessly sculpt their queries across all attributes (see Fig. 6.2).

In terms of specific interaction techniques for multivariate graphs, GraphDice supports the following:

- **Smoothly changing visual mapping:** The key feature of data space navigation [18] is to smoothly change the mapping of attribute dimensions to positional (X and Y) visual variables using an animated transition. GraphDice does not discriminate between intrinsic and computed attributes, thereby allowing the analyst to transition from a geographic or computed graph layout to other attributes such as degree, centrality, age, gender, income, etc.
- **Pivoting:** Data space navigation in GraphDice also allows for pivoting a multivariate graph to study different slices, or *facets*, of the data. This interaction is inspired by PivotGraphs [69], and also incorporates node and link aggregation to minimize overplotting and to summarize a large number of data points. Similarly, GraphDice also summarizes multiple time points into intervals that are visible during pivoting.
- **Query sculpting:** Query sculpting is a faceted filtering technique that is closely integrated with the data space navigation and pivoting functionality in GraphDice. The analyst can use lasso, bounding box, or interval selection on the main node-link display to create queries in the dataset. These queries are maintained in a query control box, which also summarizes the size, distribution, and name of each query. Analysts can then use data space navigation to pivot the query, allowing them to sculpt it by adding additional constraints on other attribute dimensions.

6.3.2 GraphTrail

The GraphTrail [14] visual analytics tool by Dunne et al. supports exploration of graph data where the nodes and links are both multivariate—containing multiple attributes, as already prominently discussed in this chapter—as well as multimodal (called *heterogeneous* in the paper)—where the nodes or links are of different types, or *modes*. The work presents two case studies: (1) publication data for the ACM CHI conference (Fig. 6.3), where nodes contain attributes such as year, title, name, locations, and date, and the modes are authors, papers, and proceedings; and (2) a large-scale archeological graph of artifacts consisting of 24 different node modes and 35 link modes. The GraphTrail tool supports the following specific interaction techniques for multivariate graphs:

- **Aggregation:** The tool presents aggregated views of graphs in self-contained charts such as bar charts, tag clouds, and tables instead of the raw graph data as a traditional node-link diagram. The purpose is to use familiar and readable visual summaries as opposed to the full graph dataset.

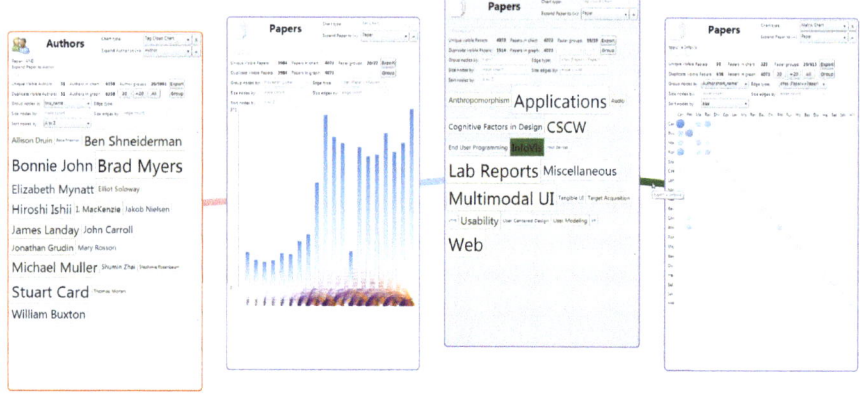

Fig. 6.3. GraphTrail [14] overview of a multivariate co-authorship dataset for the ACM CHI conference. The screenshot shows examples of the tag cloud, hybrid bar chart, and matrix chart supported by the tool.

- **Visual history:** While not strictly a multivariate graph interaction technique, GraphTrail provides an innate visual interaction history by maintaining each exploration branch as a chain, or trail, of connected charts. This allows the analyst to refer back to the exploration path, which may potentially be branching, at any time.
- **Exploratory interactions:** The tool supports three specific interaction techniques for multivariate graph exploration:

    Filtering and merging: Selecting subsets of a dataset for drill-down and merging disparate subsets into a single chart using direct manipulation.

    Pivoting: Transitioning between different edge and node types (i.e., modes) to explore multimodal relationship in the graph.

    Cloning: Duplicating subsets and charts with dependencies to avoid having to propagate upstream changes to connected child charts.

6.3.3 State Transition Networks

Pretorius and van Wijk [52] present a multivariate graph visualization technique for visual inspection of state transition graphs (Fig. 6.4). Such graphs are common for complex systems and are often used for design, debugging, and evaluation. The visual representation is based on separating the different modes in these state transition graphs and showing two modes at a time, essentially as a bipartite graph layout with nodes on each side of the display and the links (and edge labels) connecting the modes in-between. The implementation allows the user to navigate between which two modes they

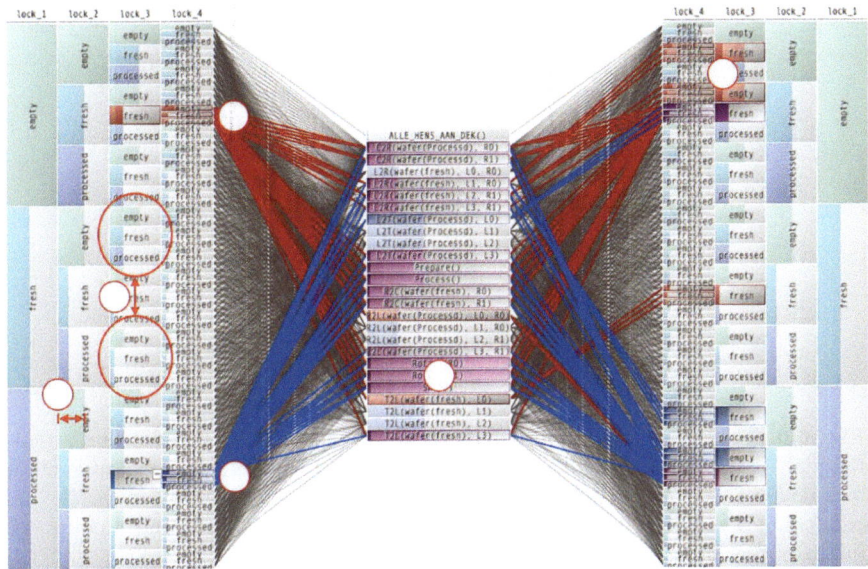

Fig. 6.4. A multivariate graph visualization designed by Pretorius and Van Wijk [52] of a state transition graph created by a system analyst

wish to drill down into. In addition, the system supports several dedicated multivariate graph interaction techniques:

- **Selection and highlighting:** A simple but key interaction technique is the ability to select a node (or a cluster of nodes) in the visual representation, causing all contained or connected nodes and edges to be highlighted in red. This interaction is coupled with appropriate visual representations that also highlight multivariate attributes in the connecting edges.
- **Filtering:** In conjunction with the selection technique, the multivariate graph implementation also supports adding to or subtracting from the selection to further refine the analyst's exploration.
- **Clustering:** To cope with the large scale of the state transition graphs, the prototype implementation supports clustering and aggregating nodes and edges based on attributes and labels.

6.3.4 Parallel Node-Link Bands

Similar to GraphTrail, the parallel node-link bands (PNLBs) [22] method is a graph visualization technique for multimodal and multivariate graphs, i.e., graphs where nodes and links not only have multiple attributes, but also belong to two or more different modes, or types. However, instead of focusing on aggregated charts summarizing the network, PNLBs draw on the work by Pretorius and van Wijk [52] (discussed above) to retain the node-link

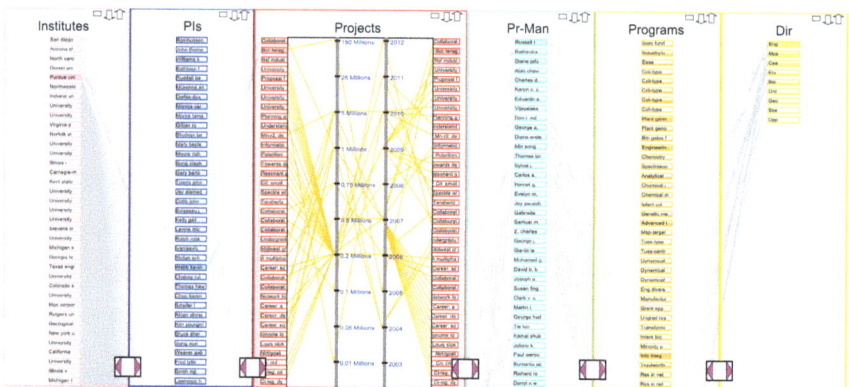

Fig. 6.5. The parallel node-link bands (PNLBs) [22] technique visualizing a multimodal NSF funding graph. The parallel coordinate inset is a specific interaction technique called "open sesame" for drilling down into one or several scalar attributes of a set of nodes; in this case, it is visualizing the year and amount for funded projects.

visual metaphor but separates the nodes by their respective mode into specific *bands* organized by slicing the viewport vertically (Fig. 6.5). Unlike Pretorius and van Wijk, PNLBs generalize to any number of bands, although they only show links between adjacent bands, suppressing all other link modes to minimize visual clutter. For this reason, the technique also borrows many ideas from semantic substrates [57], where node modes are organized into spatially disjoint substrates. However, PNLBs were designed specifically for multivariate graph exploration, and provide the following interaction techniques to support this goal:

- **View distortion:** PNLB bands can be zoomed and panned independently of each other; furthermore, they can also be designed to support semantic zooming. One particular use of this is to enable view distortion where a fisheye function around a selection or the user's mouse cursor can smoothly expand and compress the visual marks representing the nodes in the graph.
- **Multivariate drill-down:** The tool allows the analyst to drill down into entire bands or individual nodes to uncover the multivariate attributes "hidden" in the data. For example, tag cloud and details-on-demand popups can show summaries or the full details of a node. Furthermore, a specific interaction technique dubbed "open sesame" (visible in Fig. 6.5) integrates a parallel coordinate inset within an expanded node band to show quantitative data for those nodes; parallel coordinates were chosen because they closely mimic the overall visual design of PNLBs, but other chart types can be integrated as well.
- **Multimodal drill-down:** Another drill-down option focuses on the topological nature of the graph by exposing the within-network relations

within the dataset, i.e., the links that connected nodes of the same mode. This is a necessary mechanism since the PNLBs technique is designed to primarily show between-mode links for adjacent mode bands.

6.4 Recommendations and Guidelines

When designing or evaluating interactive visualizations for multivariate network data, it is useful to consider potential interaction techniques with regard to their usability. Indeed, this is even more important when testing novel unproven interaction techniques. There exists a large body of experience in the Human-Computer Interaction community regarding the usability of interaction techniques. In this section, we describe some well accepted usability principles and interaction design guidelines and discuss them within the context of multivariate network visualization. The information in this section should provide a useful lens through which to assess the appropriateness of particular interaction techniques.

We group these guidelines into three broad categories—Learnability, Flexibility and Robustness—as suggested by Dix et al. [12]. We also draw from other sources, including Cognitive Dimensions of Notations [8, 25], which offers some useful vocabulary for discussing design and choice of interactions as well as evaluating the impact a design will have on users.

6.4.1 Learnability

Learnability describes a set of principles that can be used to determine the ease with which a new user can begin productive work with the system [12]. This is especially important when designing interactions for multivariate network visualization since these will often present a large amount of data, and complex interfaces for exploring the dimensions of the data. Hence, anything that can be done to help users quickly learn the system and accompanying interactions is crucial.

One important aspect of Learnability is **Predictability**, which simply states that a given interaction should always behave predictably, i.e., exhibit deterministic behavior. Also, it should exhibit **Consistency** in that an interaction that can be performed on one element of the visualization should be able to be applied to other objects and produce similar results. This means sticking with established conventions for network and multivariate visualization, such as those described earlier in this chapter.

Other aspects of Learnability are **Familiarity** and **Generalizability** which deal with creating interfaces and interactions that map as closely as possible to the real world or similar interfaces the user will already be familiar with. Ideally, this is done to maximize utilization of users' past experience. This includes making use of familiar metaphors where applicable, as discussed at length in Chap. 7.

Affordance [48] describes the ability of physical or digital objects to suggest how they may be interacted with through their appearance. For example, the handles on a drawer afford the user the ability to pull out the drawer. Similarly, the appearance of standard GUI controls like sliders and buttons suggest how users may interact with them.

When designing novel interfaces for working with multivariate network visualizations it might not always be possible to give interactive controls or elements these obvious affordances due to the complexity of the interface, but the appearance of controls can still sometimes be enhanced in subtle ways to hint at their intended use, or at least at the possibly for interaction. For example, with colored hyperlinks in web documents, it is not always obvious what effect clicking a link with have, but the user knows it will do something and can make an educated guess based on the link text and surrounding context. Affordance is related to the discoverability of interactive capabilities in the interface; with the profusion of graphical entities shown by visualization, it is always tempting to provide interactions contextual to specific entities, but without affordance, the chance that a novice user will discover it by chance is very low. When designing user interface components or assigning contextual interactions to graphical entities, ask yourself if they adequately express their role to the user?

Avoiding **Hidden Dependencies** [8, 25] means that the link between connected items should be visible and obvious to the user. This can be a problem in any system with filtering and search where the visualization may just show a subset of the results. Ideally, if the information linking matching elements is significant the visualization can show the smallest subset of the network that connects these nodes and edges in the search results. This becomes a greater problem when an interaction leads to a surprising result due to such a hidden dependency.

Supporting **Progressive Evaluation** [8, 25] means allowing users to be able to take a break at any time and take stock of their progress so far. This is especially important for exploratory tasks involving novice users.

6.4.2 Flexibility

Flexibility groups a set of principles that deal with best practices for the avenues for information exchange between the user and visualization system.

Cognitive load [45] is based on the fact that humans can hold relatively little information in short term memory—famously, seven plus/minus two pieces of information. As a result, we must consider the amount of information that needs to be retained in working memory in order to effectively work with a system or interface. As much as possible there is a need to alleviate the user having to commit unnecessary information to memory. In terms of visualizations of multivariate networks, this means relevant data should be highly visible and the interface should make clear what data and attributes the users are looking at, how they reached this point, as well as how they can return to earlier points of exploration. It should be possible to delegate the task of remembering information for complex processing tasks to the system where these details should be presented in a fashion that is easy for the users to understand.

Cognitive load and other limits of visualization are discussed in more detail in the chapter on scalability considerations (Sect. 10.1).

Fitts's law [21] is a model that describes the act of pointing. It says that the time taken to rapidly move a pointing device to a target object is proportional to the size of the target and the distance to it. The implication of this for interaction and interface design is that the most frequently used controls should be the closest and largest. In the case of visualizations for multivariate networks, a prerequisite for answering the question of where to put the controls would be understanding the kinds of tasks that users were going to perform most commonly with the visualization, since context matters [1].

Visualizations should have low **Viscosity** [8, 25], i.e., common tasks should be able to be accomplished with a minimal number of actions or effort on the part of the users.

Abstraction [8, 25] involves providing shortcuts to the user in order to facilitate them working efficiently with logical sets of the data at once. This is often vital in multivariate network data, since the aim is to allow the user to manipulate the visualization at the level of a particular attribute or dimensions of the data rather than forcing them to interrogate the properties associated with individual nodes and edges themselves. Abstractions should be used where possible, since these simplify many tasks and help with understanding the network and associated variables.

Terseness and **Diffuseness** [8, 25] state that it is important to dedicate appropriate amounts of display space to the various elements of the visualization, rather than devoting too much or too little space to them. This may seem obvious, but you should think about and question the space that is being used to show various elements of the data set.

An important general quality to strive for in designing interactive visualizations of multivariate networks is providing good **Guidance**, both in terms of dimensions and graph structure. That is, when the user can not currently see some particular dimension of the data or a section of network, we would like to let them know this information exists and also give them some estimation of the importance of the non-visible information.

6.4.3 Robustness

Robustness principles relate to how the system supports the user in accomplishing their goals.

Direct Manipulation [55] describes interaction that is performed directly on objects and provides continuous, fast feedback in response to change. Another way of thinking about this is providing **continuity** and thus avoiding abrupt changes that could potentially confuse the user. When an action is not being performed by the user directly, it can sometimes be smoothly animated to achieve a similar effect. Direct manipulation interactions map more closely to object behavior in the real world and thus has a few important benefits; it allows users to quickly determine or predict the final outcome of their action, it allows them to more easily realize when performing an action would lead to unintended consequences, and to more easily reverse an action. It can also allows users to reach a desired state in a single action that would otherwise require several actions when only seeing incremental effects of the system after each individual action.

Recoverability [12] is an important robustness principle for most user interaction. It suggests that our visualization systems should easily allow the user to undo any action they have made in error. As we mentioned before, **responsive** interactions such as **direct manipulation** approaches also help in this regard.

Premature Commitment [8, 25] means users should not have to make any decision before they have adequate information to base it upon. This can often be solved by providing a flexibility to the system where the user can reach a particular result or view of the data via multiple paths, rather than just a single specific sequence of actions. Also, the user should have the ability to try things out without committing to them (**Provisionality**).

Error-Proneness [8, 25] describes the ability of the system to induce errors from the user and not protect them from these mistakes. When evaluating an interface to a complicated multivariate network visualization we must consider what is being shown to the user. Could they easily mistake or confuse some aspect of the visualization and reach an incorrect inference or conclusion? We want to avoid this.

Finally, it should be noted that creating effective interactive visualizations for multivariate network data is a difficult task that combines the inherent complexity of navigating large graphs with understanding and exploring multiple dimensions. As noted by Pike et al. [51], it is important to remember that the purpose of interaction is to enable an analytic discourse during which users build, test and refine knowledge. Hence, the design of new interaction approaches should carefully focus on the likely aims, intentions and actions of the user. Designs should also be formally evaluated through user studies and have their effectiveness tested with real users.

6.5 Challenges and Vision

This chapter has discussed and classified various interactive techniques for multivariate and network visualization. It has explored their use in several effective multivariate network visualization systems, and has described guidelines for designing successful interaction approaches. Here we conclude the chapter by outlining what we see as the major unsolved challenges in this space, and offering our thoughts on where research in this field might head in coming years.

We describe several challenges, which we broadly group into data type, data exploration, user interfaces and evaluation categories.

Data Characteristics

Scalability: Dealing with scalability is obviously a major challenge. While interaction can help with some of the issues, it is often complicated by scale. For example, can we fully show both dimensions and network data, or must we reduce the complexity of the data and its presentation? Is interaction in real time still possible on large data sets? Scalability considerations for multivariate network visualization are be discussed in detail in Chap. 10.

Temporal Data: Another challenge is in dealing with multivariate networks with a temporal dimension. Given that humans have a particular understanding of time, it can be useful to leverage this familiarity and treat it specially when designing interactive visualizations. This is explored in detail in Chap. 8.

Data Exploration and Comprehension

Understanding the Information Landscape: The first big challenge we see is that visualization research and systems often give good individual detailed views of particular facets of the data, but don't necessarily offer a visual interface that allows understanding of the entire information landscape at a high level. The difficulty is in giving enough of a sense of what the data is, conveying its meaning, as well as hinting at the dimensions or facets of the data that could be worth exploring in greater depth.

In connection to this, there needs to be more work on automatic and semi-automatic identification of important nodes and dimensions for directing the user during analysis tasks, as well as subsequent evaluation of such approaches.

Provenance of Exploration Process: We think a significant unaddressed challenge is supporting provenance in multivariate network visualizations. This involves assisting the user in tracking the exploration process, and the history of their interactions with the system. This is important since these actions form a critical part of the analysis process. While the information visualization community is aware of the importance of recording and showing which interactions were performed and when [24, 49, 51, 65], these methods either do not support multivariate networks or only to a small extent (e.g., selecting variables for visual mapping and search).

Building Interactive User Interfaces

Ad-hoc Design of User Interfaces: Currently, many multivariate network visualizations are built by extending upon existing systems using traditional interfaces. This is generally not desirable when the interfaces are required to encapsulate so much complexity. Ideally, we would like to see more approaches utilising principled design from the beginning. That is, designing interfaces and interactions specifically to support required application rather than just bolting additional controls and complexity onto existing interfaces. Specifically, this means techniques should be designed based on principles and guidelines like those presented in Sect. 6.4, but also be given formal user testing to prove their effectiveness.

In the case where multivariate network visualizations make use of **multiple views** there is a challenge in providing elegant and simple mechanisms to manage them. This is important since a single view will never be adequate to explore large multivariate networks. Users will always spend significant time controlling, comparing and navigating between views. Additionally, particular application domains often require their own unique multiple view configurations that are still a challenge to build interactively [70].

This is not to say that we can't have a set of general purpose visualization components and reuse them, but in order for such objects to be useful they will require standardized, consistent interaction. We need not just a common vocabulary and behavior, but also an understanding of their specificity and efficiency for particular uses.

Emerging Hardware: Multivariate network visualization could potentially make use of emerging hardware such as multi-touch tables and tablets, contact-less input devices (Microsoft Kinect, Leap Motion Controller), wall-sized displays, and even 3D stereoscopic displays or immersive cave environments. Additionally, there is the possibility of building novel gadgets or devices specifically for interacting with this sort of data and visualization.

Of course, these technologies are exciting to use or witness for the first time but there are a lot of unknowns surrounding their use. Do more pixels help solve the problems we have discussed? Can we benefit from extra dimension in 3D without the typical downsides, such as users becoming disoriented or lost? Can we utilize navigational multi-touch gestures? Can multimodal input provide extra benefits specifically for navigating multidimensional graphs? How can we support collaborative data exploration of multivariate networks between multiple simultaneous local or remote users? Should we build one visualization application for a specific piece of hardware (e.g., a multi-touch tablet) or can we design interactions that will adapt to a range of input devices and output capabilities?

The description of the input side of interactive applications is still at its infancy in HCI, and the management of powerful interaction approaches is relatively new in the visualization field. Adapting visualization interactions to the new setups is therefore a huge challenge [33–35]. This issue is beginning to get research attention. New ideas such as proximity-based interaction on high-walls [34] and new models for interactions across devices [35] are emerging. Research in this area will likely begin by exploring specific hardware configurations and progressively evolve towards some unification for classes of technologies over time.

Evaluation

There is a real need for formal **evaluation** of interaction approaches used for visualization of multivariate networks. There are a range of issues and concerns here. Firstly, there needs to be more consideration of the tasks for which the visualizations are to be used. Do we know what users really need when dealing with multivariate networks? How does this change when they are exploring vs. checking a hypothesis vs. using the visualization to convince another person of some fact? Chapter 5 provides an overview of tasks connected to analysis of multivariate networks, in particular.

Specifically for interaction, we need to evaluate additional factors. For example, when is interaction supportive and when can it become a burden? We need more evaluation of techniques supporting exploration and offering guidance so that the users do not get "lost" in the data space.

Evaluation results and experience partly exist for networks. However, the approaches and studies do not explicitly include and deal with multivariate networks. We need to enhance our understanding around faceted exploration along both multivariate and network data at the same time, since much of what we know applies only to exploration within a single dimension.

Acknowledgements

The genesis and planning of this chapter took place at the Dagstuhl Seminar #13201 "Information Visualization – Towards Multivariate Network Visualization" held in May 2013. We wish to thank Guy Melançon (Université Bordeaux 1, Bordeaux, France) and Robert Kosara (Tableau Software, United States) for their useful contributions during these discussions. Michael Wybrow was supported by the Australian Research Council (ARC) Discovery Project grant DP110101390.

References

1. Appert, C., Beaudouin-Lafon, M., Mackay, W.E.: Context matters: Evaluating interaction techniques with the CIS model. In: Fincher, S., Markopoulos, P., Moore, D., Ruddle, R. (eds.) People and Computers XVIII Design for Life, pp. 279–295. Springer, London (2005)
2. Archambault, D., Munzner, T., Auber, D.: GrouseFlocks: Steerable exploration of graph hierarchy space. IEEE Transactions on Visualization and Computer Graphics 14(4), 900–913 (2008)
3. Auber, D., Archambault, D., Bourqui, R., Lambert, A., Mathiaut, M., Mary, P., Delest, M., Dubois, J.: Melançon, G.: The Tulip 3 Framework: A scalable software library for information visualization applications based on relational data. Technical Report RR-7860, INRIA (January 2012)
4. Becker, R.A., Cleveland, W.S.: Brushing scatterplots. Technometrics 29(2), 127–142 (1987)
5. Bertini, E., Rigamonti, M., Lalanne, D.: Extended excentric labeling. In: Proceedings of the 11th Eurographics / IEEE – VGTC Conference on Visualization, EuroVis 2009, pp. 927–934. Eurographics Association, Aire-la-Ville (2009)

6. Bezerianos, A., Chevalier, F., Dragicevic, P., Elmqvist, N., Fekete, J.D.: GraphDice: A system for exploring multivariate social networks. Computer Graphics Forum 29(3), 863–872 (2010)
7. Bier, E.A., Stone, M.C., Pier, K., Buxton, W., DeRose, T.D.: Toolglass and magic lenses: the see-through interface. In: Proceedings of the 20th Annual Conference on Computer Graphics and Interactive Techniques, SIGGRAPH 1993, pp. 73–80. ACM, New York (1993)
8. Blackwell, A.F., Britton, C., Cox, A.L., Green, T.R.G., Gurr, C.A., Kadoda, G.F., Kutar, M., Loomes, M., Nehaniv, C.L., Petre, M., Roast, C., Roe, C., Wong, A., Young, R.M.: Cognitive dimensions of notations: Design tools for cognitive technology. In: Beynon, M., Nehaniv, C.L., Dautenhahn, K. (eds.) CT 2001. LNCS (LNAI), vol. 2117, pp. 325–341. Springer, Heidelberg (2001)
9. Buja, A., McDonald, J.A., Michalak, J., Stuetzle, W.: Interactive data visualization using focusing and linking. In: Proceedings of the 2nd Conference on Visualization, VIS 1991, pp. 156–163. IEEE Computer Society Press, Los Alamitos (1991)
10. Card, S., Mackinlay, J., Shneiderman, B.: Readings in Information Visualization: Using Vision to Think. Morgan Kaufmann Publishers (1999)
11. Collins, C., Carpendale, S.: VisLink: Revealing relationships amongst visualizations. IEEE Transactions on Visualization and Computer Graphics 13(6), 1192–1199 (2007)
12. Dix, A., Finlay, J., Abowd, G.D., Beale, R.: Human-Computer Interaction, 3rd edn. Prentice-Hall, Inc. (2003)
13. Dörk, M., Riche, N.H., Ramos, G., Dumais, S.: PivotPaths: Strolling through faceted information spaces. IEEE Transactions on Visualization and Computer Graphics 18(12), 2709–2718 (2012)
14. Dunne, C., Riche, N.H., Lee, B., Metoyer, R., Robertson, G.: GraphTrail: Analyzing large multivariate, heterogeneous networks while supporting exploration history. In: Proceedings of the ACM Conference on Human Factors in Computer Systems, pp. 1663–1672 (2012)
15. Dwyer, T., Marriott, K., Schreiber, F., Stuckey, P., Woodward, M., Wybrow, M.: Exploration of networks using overview+detail with constraint-based cooperative layout. IEEE Transactions on Visualization and Computer Graphics 14(6), 1293–1300 (2008)
16. Dwyer, T., Marriott, K., Wybrow, M.: Dunnart: A constraint-based network diagram authoring tool. In: Tollis, I.G., Patrignani, M. (eds.) GD 2008. LNCS, vol. 5417, pp. 420–431. Springer, Heidelberg (2009)
17. Elmqvist, N., Do, T.N., Goodell, H., Henry, N., Fekete, J.D.: ZAME: Interactive large-scale graph visualization. In: Proceedings of the IEEE Pacific Symposium on Visualization, pp. 215–222 (2008)
18. Elmqvist, N., Dragicevic, P., Fekete, J.D.: Rolling the dice: Multidimensional visual exploration using scatterplot matrix navigation. IEEE Transactions on Visualization and Computer Graphics 14(6), 1141–1148 (2008)
19. Elmqvist, N., Fekete, J.D.: Hierarchical aggregation for information visualization: Overview, techniques, and design guidelines. IEEE Transactions on Visualization and Computer Graphics 16(3), 439–454 (2010)
20. Fekete, J.D., Plaisant, C.: Excentric labeling: dynamic neighborhood labeling for data visualization. In: Proceedings of the SIGCHI Conference on Human Factors in Computing Systems, CHI 1999, pp. 512–519. ACM, New York (1999)
21. Fitts, P.M., Peterson, J.R.: Information capacity of discrete motor responses. Journal of Experimental Psychology 67, 103–112 (1964)

22. Ghani, S., Kwon, B.C., Lee, S., Yi, J.S., Elmqvist, N.: Visual analytics for multimodal social network analysis: A design study with social scientists. IEEE Transactions on Visualization and Computer Graphics 19(12), 2032–2041 (2013)
23. Gladisch, S., Schumann, H., Tominski, C.: Navigation recommendations for exploring hierarchical graphs. In: 9th International Symposium on Visual Computing (ISVC 2013). Advances in Visual Computing, pp. 36–47. Springer (2013)
24. Gotz, D., Zhou, M.X.: Characterizing users' visual analytic activity for insight provenance. Information Visualization 8(1), 42–55 (2009)
25. Green, T.R.G.: Cognitive dimensions of notations. In: Sutcliffe, A., Macaulay, L. (eds.) Proceedings of the 5th Conference of the British Computer Society, Human-Computer Interaction Specialist Group - People and Computers V, pp. 443–460. Cambridge University Press, New York (1989)
26. Hachul, S., Jünger, M.: Large-graph layout algorithms at work: An experimental study. Journal of Graph Algorithms and Applications 11(2), 345–369 (2007)
27. van Ham, F., Perer, A.: Search, Show Context, Expand on Demand: Supporting large graph exploration with degree-of-interest. IEEE Transactions on Visualization and Computer Graphics 15(6), 953–960 (2009)
28. Heer, J., Boyd, D.: Vizster: Visualizing online social networks. In: IEEE Symposium on Information Visualization, pp. 32–39 (2005)
29. Heer, J., Card, S.K.: DOITrees revisited: scalable, space-constrained visualization of hierarchical data. In: Proceedings of the ACM Conference on Advanced Visual Interfaces, pp. 421–424 (2004)
30. Heer, J., Shneiderman, B.: Interactive dynamics for visual analysis. Commun. ACM 55(4), 45–54 (2012)
31. Henry, N., Fekete, J.D.: MatrixExplorer: a dual-representation system to explore social networks. IEEE Transactions on Visualization and Computer Graphics 12(5), 677–684 (2006)
32. Henry, N., Fekete, J.D., McGuffin, M.J.: NodeTrix: a hybrid visualization of social networks. IEEE Transactions on Visualization and Computer Graphics 13(6), 1302–1309 (2007)
33. Isenberg, P., Carpendale, S., Bezerianos, A., Henry, N., Fekete, J.D.: CoCoNutTrix: collaborative retrofitting for information visualization. IEEE Comput. Graph. Appl. 29(5), 44–57 (2009)
34. Jakobsen, M.R., Sahlemariam Haile, Y., Knudsen, S., Hornbaek, K.: Information visualization and proxemics: Design opportunities and empirical findings. IEEE Transactions on Visualization and Computer Graphics 19(12), 2386–2395 (2013)
35. Jansen, Y., Dragicevic, P.: An interaction model for visualizations beyond the desktop. IEEE Transactions on Visualization and Computer Graphics 19(12), 2396–2405 (2013)
36. Jusufi, I., Dingjie, Y., Kerren, A.: The Network Lens: Interactive exploration of multivariate networks using visual filtering. In: Proceedings of the 14th International Conference on Information Visualisation (IV 2010). pp. 35–42. IEEE Computer Society (2010)
37. Jusufi, I., Kerren, A., Zimmer, B.: Multivariate network exploration with JauntyNets. In: Proceedings of the 17th International Conference on Information Visualisation (IV 2013), pp. 19–27. IEEE Computer Society Press (2013)

38. Kadivar, N., Chen, V., Dunsmuir, D., Lee, E., Qian, C., Dill, J., Shaw, C., Woodbury, R.: Capturing and supporting the analysis process. In: IEEE Symposium on Visual Analytics Science and Technology, VAST 2009, pp. 131–138 (2009)
39. Kerren, A., Schreiber, F.: Toward the role of interaction in visual analytics. In: Proceedings of the 2012 Winter Simulation Conference (WSC 2012), pp. 420:1–420:13. IEEE Computer Society Press (2012)
40. Kruskal, J., Wish, M.: Multidimensional Scaling, Sage University papers, vol. 11(07). SAGE Publications (1978)
41. Lima, M.: Visual Complexity: Mapping Patterns of Information. Princeton Architectural Press (2011)
42. Liu, Z., Navathe, S.B., Stasko, J.T.: Network-based visual analysis of tabular data. In: IEEE VAST, pp. 41–50 (2011)
43. May, T., Steiger, M., Davey, J., Kohlhammer, J.: Using signposts for navigation in large graphs. Computer Graphics Forum 31(3), 985–994 (2012)
44. McGuffin, M.J., Jurisica, I.: Interaction techniques for selecting and manipulating subgraphs in network visualizations. IEEE Transactions on Visualization and Computer Graphics 15(6), 937–944 (2009)
45. Miller, G.A.: The magical number seven, plus or minus two: some limits on our capacity for processing information. Psychological Review 63, 81–97 (1956)
46. Moscovich, T., Chevalier, F., Henry, N., Pietriga, E., Fekete, J.D.: Topology-aware navigation in large networks. In: Proceedings of the ACM Conference on Human Factors in Computing Systems, pp. 2319–2328 (2009)
47. Munzner, T., Guimbretière, F., Tasiran, S., Zhang, L., Zhou, Y.: TreeJuxtaposer: scalable tree comparison using focus+context with guaranteed visibility. In: ACM SIGGRAPH 2003 Papers, SIGGRAPH 2003, pp. 453–462. ACM (2003)
48. Norman, D.A.: The design of everyday things. Basic Books (2002)
49. North, C., Chang, R., Endert, A., Dou, W., May, R., Pike, B., Fink, G.: Analytic provenance: process+interaction+insight. In: Extended Abstracts of the ACM Conference on Human Factors in Computing Systems, pp. 33–36 (2011)
50. Perlin, K., Fox, D.: Pad: an alternative approach to the computer interface. In: Proceedings of the 20th Annual Conference on Computer Graphics and Interactive Techniques, SIGGRAPH 1993, pp. 57–64. ACM, New York (1993)
51. Pike, W.A., Stasko, J., Chang, R., O'Connell, T.A.: The science of interaction. Information Visualization 8(4), 263–274 (2009)
52. Pretorius, A.J., van Wijk, J.J.: Visual inspection of multivariate graphs. In: Proceedings of the Eurographics / IEEE – VGTC Conference on Visualization, pp. 967–974 (2008)
53. Robinson, A.H.: The thematic maps of Charles Joseph Minard. Imago Mundi 21, 95–108 (1967)
54. Roth, R.E.: An empirically-derived taxonomy of interaction primitives for interactive cartography and geovisualization. IEEE Transactions on Visualization and Computer Graphics 19(12), 2356–2365 (2013)
55. Shneiderman, B.: Direct manipulation: A step beyond programming languages. IEEE Computer 16(8), 57–69 (1983)
56. Shneiderman, B.: The eyes have it: a task by data type taxonomy for information visualizations. In: Proc. Int. Symp. on Visual Languages, pp. 336–343 (1996)
57. Shneiderman, B., Aris, A.: Network visualization by semantic substrates. IEEE Transactions on Visualization and Computer Graphics 12(5), 733–740 (2006)

58. Shneiderman, B., Plaisant, C.: Designing the User Interface – Strategies for Effective Human-Computer Interaction, 5th edn. Addison-Wesley (2010)
59. Silva, C., Freire, J., Callahan, S.: Provenance for visualizations: Reproducibility and beyond. Computing in Science Engineering 9(5), 82–89 (2007)
60. Smoot, M.E., Ono, K., Ruscheinski, J., Wang, P.L., Ideker, T.: Cytoscape 2.8: new features for data integration and network visualization. Bioinformatics 27(3), 431–432 (2011)
61. Spenke, M., Beilken, C., Berlage, T.: FOCUS: the interactive table for product comparison and selection. In: Proceedings of the 9th Annual ACM Symposium on User Interface Software and Technology, UIST 1996, pp. 41–50. ACM, New York (1996)
62. Tekušová, T., Kohlhammer, J.: Visual analysis and exploration of complex corporate shareholder networks. In: Proceedings of the SPIE Conference on Visualization and Data Analysis (VDA 2008), p. 68090F. International Society for Optics and Photonics (2008)
63. Tominski, C., Abello, J., van Ham, F., Schumann, H.: Fisheye tree views and lenses for graph visualization. In: Proceedings of the Internationl Conference on Information Visualization, pp. 17–24 (2006)
64. Tu, Y., Shen, H.W.: Balloon focus: a seamless multi-focus+ context method for treemaps. IEEE Transactions on Visualization and Computer Graphics 14(6), 1157–1164 (2008)
65. von Landesberger, T., Fiebig, S., Bremm, S., Kuijper, A., Fellner, D.: Interaction taxonomy for tracking of user actions in visual analytics applications. In: Huang, W. (ed.) Handbook of Human-Centric Visualization, pp. 653–670. Springer (2014)
66. von Landesberger, T., Bremm, S., Bernard, J., Schreck, T.: Smart query definition for content-based search in large sets of graphs. In: Proceedings of EuroVAST, pp. 7–12. European Association for Computer Graphics (Eurographics), Eurographics Association, Goslar (2010)
67. von Landesberger, T., Görner, M., Rehner, R., Schreck, T.: A system for interactive visual analysis of large graphs using motifs in graph editing and aggregation. In: Proceedings of the Vision Modeling Visualization Workshop, pp. 331–339 (2009)
68. von Landesberger, T., Kuijper, A., Schreck, T., Kohlhammer, J., van Wijk, J., Fekete, J.D., Fellner, D.W.: Visual analysis of large graphs: State-of-the-art and future research challenges. Computer Graphics Forum 30(6), 1719–1749 (2011)
69. Wattenberg, M.: Visual exploration of multivariate graphs. In: Proceedings of the ACM Conference on Human Factors in Computing Systems, pp. 811–819 (2006)
70. Weaver, C.: Building highly-coordinated visualizations in Improvise. In: IEEE Symposium on Information Visualization, pp. 159–166 (2004)
71. Wong, N., Carpendale, S., Greenberg, S.: Edgelens: An interactive method for managing edge congestion in graphs. In: Proceedings of the IEEE Symposium on Information Visualization (InfoVis 2003), pp. 51–58. IEEE (2003)
72. Yi, J.S., Kang, Y.A., Stasko, J.T., Jacko, J.A.: Toward a deeper understanding of the role of interaction in information visualization. IEEE Transactions on Visualization and Computer Graphics 13(6), 1224–1231 (2006)

7
Novel Visual Metaphors for Multivariate Networks

Jonathan C. Roberts, Jing Yang, Oliver Kohlbacher, Matthew O. Ward, and Michelle X. Zhou

As visualization researchers we are often in search of new designs. In particular, when the data is huge, and there are many variables, it is challenging for the developer to imagine new designs that would be effective. As well as imagining a new visual projection methodology, developers need to create designs that enable users to explore, interact and perceive the data. While this design challenge is a broad issue in the subject of data visualization, multivariate network data offers specific challenges to the developer and designer. Users wish to understand network data that contains many nodes and edges, with many variables at each node and on each edge. In fact, the graph visualizations that are often used with this type of data contain many thousands of nodes and edges and are complex to understand.

Consequently, traditional visualization methods soon breakdown. One solution is to represent the data as a three-dimensional network. For instance, colored spheres are located in a three-dimensional space, that are connected to each other by straight lines. However these types of visualizations are ineffective, because they contain much occlusion. So it is therefore difficult for a user to visually understand routes or connections that are located within the networks. Certainly many clever clustering algorithms have been created, but it is still demanding for users to perceive clusters in these huge and complex networks.

Because of these challenges, users of these graphs often name them 'birds nest', 'hairballs', 'cat hairs', or 'balls of yarn'. They are a tangled and complicated mess of lines and nodes. Labels form another challenge. Label everything, and the screen becomes a mess of overlapping text, while label nothing and the data could be meaningless.

All these issues provide a huge challenge for visualization designers. They also offer a massive opportunity for designers to think differently. This leads us to ponder many questions. How can developers create novel depictions of multivariate network data? What design processes can be used to guide the development of new visualization solutions? Where can inspiration come from?

In particular, inspiration for *novel design ideas* can come from a variety of sources. Psychologists and designers have had many discussions over the *ideation*[1] process. Their goals are to make the process more prescriptive and predictable. Indeed, while these processes do provide a general framework, there is no straightforward solution that will guarantee success. In addition, their processes are universal: they are general processes that can be used to help develop new design ideas on a range of topics.

Our goal, however, is to help develop new ideas for data visualization, specifically for multivariate network data. Our ideas were initiated during discussions that were held in the Dagstuhl Seminar #13201 [22] that resulted in this book. Through sharing our experiences over the creation of novel designs for multivariate network visualization, and from our knowledge of visualizing multivariate network data, we realized that there are several principles and concepts that can help designers. In fact, we noticed that there were several inspirational designs that were *metaphorical*. We therefore gathered and collated these concepts from our own experience of crafting solutions for multivariate network visualization and also from the literature.

Consequently our hypothesis is that new ideas can be inspired by transferring ideas that occur in one domain to another, i.e., they can be formed by looking at other 'things'. These design ideas are *metaphors*. Often these inspirational concepts are constructs; they are structures that we can 'hook' our visualization ideology onto.

In this chapter we focus on metaphors for the visualization of multivariate network data. These are 'ideas', where aspects of the concepts are *characteristics* of something else. They may be *derivatives* and therefore abstract, but they are derived from something. Indeed they may appear to be extremely 'far from the original', or removed from their original inspiration, or may be adaptations and therefore are not identical in every aspect to the object that brought about the inspiration.

7.1 Background

7.1.1 Semantically Rich Data

Multivariate network data is semantically rich. It not only forms a connected structure where nodes are linked to each other, but the nodes have several attributes. This means that the network can be displayed in different configurations, i.e., there are several valid arrangements of the data. This not only brings forth an opportunity for creativity, but also creates a design challenge (i.e., how to display different configurations). We encourage the reader to view the tasks chapter (5) for more information on tasks with multivariate network data.

---

[1] Ideation is the process of 'idea generation'.

There is an interesting phenomenon that is highlighted through prior work. Most of the novel metaphors are designed for specific applications. For example, PeopleGarden [48] and Chat Circles [44] target on-line interaction environments such as web-based message boards and chat rooms; Thread Arcs [21] was developed for email visualization; NetLens [19] and CiteVis [39] aim to visualize content-actor networks such as citation networks. This phenomenon suggests that the ideation of novel metaphors is often *task-driven*.

The different views certainly enable several *tasks* to be completed by the user. Chapter 5 on tasks for multivariate network visualization expands these features. In that chapter, Pretorius, Purchase and Stasko highlight that users can locate, distinguish, categorize, cluster, (analyze) distributions, rank values, compare different parts of the network, associate and correlate. Investigating what tasks the user needs to perform is important. Indeed, it is clear that specific visualizations enable the user to investigate different tasks. Subsequently it may be that different metaphors enable various tasks to be performed better.

Multivariate data is found in several fields. While in this chapter we do not focus on specific data types, rather look across the subject domains, at multivariate network visualization solutions that utilize a metaphoric approach, we refer the reader specifically to related chapters that do investigate specific domains: Chap. 2 where solutions to visualize multivariate network data in software engineering are discussed, and Chap. 4 where multivariate bioinformatics network data is explored, and Chap. 3 that focuses on social media data.

7.1.2 Where Ideas Come From?

Inspiration for a particular design can come from many sources. For instance, Johnson [17] mentions that ideas develop by looking at lots of other ideas. It is through looking at many different concepts that better ideas are formed. In fact, inspiration can come from nature; consequently, bio-inspiration, or bio-mimicry is a popular topic. Taking ideas from nature is commonly used by designers.

Let's work through an example. Consider a tree. A tree has a trunk and off this main upright big branches are formed. These in turn generate smaller branches then smaller branches still, and finally leaves. This gives inspiration to many ideas in computing. It is a hierarchy. It is a natural structure. It has many facets. Subsequently, there are many tree-like objects in computing in general, and visualization specifically. For example, binary trees as data-structures [34], quad-trees in computer graphics [34], Card's cone-tree hierarchical visualization technique [32], or the treemaps [37] visualization method. Whether the visualization designer explicitly thought of trees when they were designing their visualization is a matter for discussion. However trees are ubiquitous, and would have been in the subconscious mind, whether or not they provided a direct inspiration.

Let us take the 'tree' metaphor further. Trees are not just *structural* phenomena, there are many other aspects of trees as a *metaphor* that could provide inspiration. For instance, trees are large and long-lived plants that compete for light, they drop seeds that then can grow into other trees, and many trees often grow together into a forest. The forest has an ecosystem and therefore other 'life' depends on the trees. On the ground, fallen leaves and decaying branches provide habitat, while trees give protection for animals and wind protection for humans and buildings. Fungi and bacteria also rely upon trees. They provide food for animals, their sap can be tapped for food and made into products, and the trees have many different parts: roots, branches, bark, leaves, and so on. Trees don't grow constantly throughout the year. They often drop their leaves in the fall and grow new buds in spring; the speed of growth depends on the growing conditions of that year (hence dendrochronology is possible).

So thinking deep and thinking comprehensively about a topic can engender different and more ideas. For instance, we may think further about *dependency*, where perhaps aspects of a visualization may depend on others. In fact, this happens in a parallel coordinate plot, where the perception of the data from one axis depends on it's neighbors. Perhaps we may then think how we can design a parallel coordinate plot that does not have this dependency. Take another tree-like facet: the *parts* of a tree. Roots and branches are similar, yet opposites. The root structure is a hierarchy that goes downwards in search of water, while the leaves are on the stems that grow upwards in search of light. A designer could consider a tree-like visualization in two parts. Perhaps statistical information of a population could be displayed in a tree, with female participants displayed upwards, and male downwards. This would allow comparison of male and females, and many subjects could be represented by a forest. Finally, tree rings could be used as a direct metaphor, with time progressing outward from the centre, and the width of the rings dependent on another variable.

There is a clear link between metaphors and visualization. Not only do the metaphors drive inspiration, but they are used in comprehension. Ziemkiewicz and Kosara [54] suggested that the "process of understanding a visualization therefore involves an interaction between these external visual metaphors and the user's internal knowledge representations". Their work proposes that metaphors work both ways: they both inspire, and are needed for interpretation. Their work also used trees as an exemplar metaphor.

Just by thinking about a tree in more depth, we can start to imagine how metaphors can help. In fact, this is *lateral thinking* [9]; it is not only a matter of considering the principle concept (a tree, in this case), but delving deeper into the concepts, and considering processes, connections, colors, environment, and so on, as well as the structures of these concepts. As we shall see, ideas are formed through a long study, careful consideration, joining concepts together, looking at opposites, and so on, and rarely form as a *eureka* moment.

Metaphoric concepts are not the only method of ideation. Sketching and doodling are also important techniques that are used to create and craft new visualization designs [31]. Ideas can be easily explored through sketching. In fact, merely the act of putting the ideas on paper can help the designer to hone his ideas. Sketching acts as a refinement methodology. Users can share thoughts and can easily adapt these ideas.

Even discovering solutions that don't work can help inspire solutions that do work. These unworkable 'solutions' often act as a catalyst that inspire a better result. This is similar to another concept generation method: *provocation*. Johnson [17] suggests finding *dissimilar* ideas and joining them together; through this joining up of different thoughts new ideas can be formed. de Bono, in his book "Six Thinking Hats" [8], under 'green-hat thinking', suggests choosing a random number for a page in a dictionary, and selecting a random word on that page. That random word can be used to provoke different solutions and novel ideas.

Ideas and solutions are sometimes formed from disasters, or mishaps, or errors. For example, Wiseman [47] tells the story of PostIt-note^{TM} glue being an accident, or commonly referenced as a 'solution without a problem', and the discovery of penicillin, by Alexander Fleming in 1928, was serendipitous. Each of these ideas were originally accidents that became answers to specific problems over time.

7.1.3 The Ideation Process

In one respect, design-work is a journey. Ideas come and go, some ideas *stick*, are more memorable, others are easy to understand and maybe easier to communicate. Ideas become refined by cognition and discussion with other researchers; they are honed by mulling them over and are extended through interaction with people, or through different personal experiences.

Good ideas occur by *slow* and *careful* thought over a long period of time. Johnson, in his book "Where Good Ideas Come From" [17] talks about 'slow hunches'. Often people cite a 'eureka' moment as the derivation of their idea, however, while a single instance can bring a specific idea to the fore, in reality the idea is usually something that has been considered for a while by the designer, and draws on their previous experiences and knowledge.

While each of these methods offers a solution to the thinker when he or she is stuck, we need a process to follow to create new ideas in a systematic way. Young [52] in his seminal book explains a succinct process of producing ideas: first gather materials both specific and general. Second, think, make connections and write down every idea. Third, relax. Go do something else and let your subconscious work on the problem. As soon as the idea appears, write it down. Fourth, rework the idea. Refine the concept and make it appropriate for the purpose. Expand and contract the idea to make it robust.

De Bono's [9] 'Lateral thinking' method includes many of these techniques. It consists of seven techniques: consider **alternatives** and look beyond the ideas. **Focus** on your problem and discipline your thoughts. **Challenge** yourself, and try to break away from traditional thinking. Use un-connected ideas, to provide a **random entry** to new ones. Use **provocation** statements to engender and develop new ideas. Note-take, record and journal all your ideas so as to **harvest** the ideas. Know how to **treat** your ideas and fit them into the place where they are required.

In visualization specifically a few models have been presented, including Munzner's [28] Nested Model for Visualization Design and Validation, and Roberts' Five Design Sheet (FdS) method [31]. Whatever the process that is followed, it is clear that *lateral thinking* and inspiration from metaphors are beneficial to the design process.

7.1.4 The Visual Mapping Process

The ideation process goes hand-in-hand with the any software development process. As ideas change and develop through reflection and contemplation, so does the software develop through many iterations. The metaphors and original ideas change as they are implemented into a software solution. Indeed, software is often developed through an iterative strategy. In Agile software development methodologies, mixed teams of software developers, clients and end-users, create solutions in an iterative and incremental way. This is a sensible development strategy. Not only are ideas refined when they are sketched, and written down, or through consideration, but the ideas are refined when they get developed and used by a user [31].

Particularly, the effectiveness of the metaphor may be understood when the user actually uses the tool to perform a task. By changing the metaphor in small or even major ways, it may become more effective. Consequently, it is a good strategy to get end-users involved from the very start of the process [28]. Then the created software will be better suited to the needs of the users. Users can test the solution to see if they understand the metaphor, and whether they can perceive the data correctly.

Through this *iterative* ideation and development process, there are many practical design choices that the developer (or team) need to make. For example, developers need to ascertain which structures and components of the metaphor to use? What is the task the user is to perform? How to interpret the metaphor into a visualization tool that would perform a given task? What parameters to include? What sub-structures of the metaphor to use? What retinal variables to use to depict the data? etc.

Consequently, developers need to interpret the metaphor to fit their needs. They have to translate the concepts of the metaphor into a fully functional and working visualization tool. It may be that the developer chooses a certain aspect of the metaphor for their purpose, or adapts it for their need. They also need to ascertain how different components, structures or sub-elements of the metaphor represent the data, and how it is mapped to the retinal variables. The data is then encoded by adapting the properties of the retinal variables. For instance, a person could be represented by a branch of a tree, the length of their life is encoded as the length of that branch. Children are represented by other sub branches.

As developers, we need to think about the data, how the metaphor is understood by a user, and finally how the user interacts and changes the visualization. Therefore there are several concepts: data transformation, visual transformation, and visual mapping that developers need to consider.

Data Transformation

There are many different types and configurations of datasets. Some are structured and dense, while other data could be sparse or unstructured. Add in free text fields or formatted text and it is clear that users' information varies a lot among different applications. Being aware of this variation and understanding the tasks of the user for a particular application is important. These aspects change the requirements on the data transformation.

Some tasks require summaries to be generated, where lots of data can be summarized into a few values. In this case, non-node link diagram metaphors become a natural choice for non-node-link essential information units. It is possible also to imagine that these aggregated values could be mapped to aspects of objects that take up more real-estate (for instance). Indeed, such data aggregations often occur over time, and consequently the summed-data could be visualized as rings of a tree; where there is a linear progression from the early rings (the center ones) to the external rings.

Other tasks require that individual items in the data remain intact in the visualization depiction. The data transformation therefore is 1:1. In this situation, the identity of the individual entities need to be directly passed through to the mapping process, where individual items can be mapped to individual entities of the visualization. In other data, sub-structures or sub-graphs may exist in the data. Consequently, metaphors that reflect these constructs should be designed.

For example, PeopleGarden [48] aims to help new users find appropriate on-line interaction groups to participate in and people to interact with. To achieve this goal, data portraits, namely the abstract representations of interaction histories of individual users, are identified as essential information units. Novel visual metaphors are then inspired by the data model. In particular, a flower metaphor is used to represent the data portraits, and a garden metaphor is used to combine the portraits to represent an on-line environment.

Visualization Transformation

When we select visual metaphors for representing the information units identified in the previous step, many issues need to be considered, such as: Are the metaphors familiar to the target users? If unfamiliar metaphors are used, how significant is the learning curve? Is the mapping from the information units to the metaphors consistent with common knowledge of the users? If it is inconsistent with common knowledge, to what extent will it hinder the users from effectively conducting the tasks?

For instance, Chat Circles [44] considers the social presence and activities of individuals as essential information for conveying the dynamics of on-line synchronous conversation. It therefore maps individuals to circles whose visual attributes represent their identities and activities. Moreover, a hearing range is defined for a user and only conversations of people within the hearing range are visible in the visualization. These metaphors mimic a cocktail party, which is familiar, and the mapping is consistent with common knowledge of on-line conversation users. Therefore, the resulting visualization is intuitive to the users.

Visual Mapping Transformation

Most visualizations are interactive, which means that users can interact with the visual metaphors to conduct tasks. The ideation of novel metaphors can happen when answering the following questions: How can the users manipulate the visual objects? How will the visual objects react to user input?

Wise [46] explicitly describes this method as an "ecological" approach. There is an intimate relationship between the human being and their environment; all our senses work together to gain a holistic perception of what is being viewed. This approach is inspired by Gibson's view of perception [14], where there is an intimate relationship between a human being and their environment. Some concepts *afford* certain uses. For instance, a bowl is concave and therefore it can hold water and be used to drink from. Likewise, the positions of the leaves on a tree depend on the structure of the branches, and other branches are subordinate to the trunk. So, if this structure is used for the visualization, a user may assume a similar relationship to the data. A fun example is the Collapsible Cylindrical Trees (CCT) [7]. CCT uses a telescope metaphor to produce a compact hierarchy visualization for fast navigation of web page hierarchies. The descendants of a node are represented by a cylinder. Users can pull out/collapse a cylinder into its parent cylinder to roll up/drill down the hierarchy. They can also rotate a cylinder to browse the siblings.

Indeed, the tree metaphor is a powerful design structure, and provides many levels of detail. For instance, as a user moves away from a tree, on the one hand, individual leaves are difficult to observe, but on the other hand, the overall shape of the tree can be better seen. Perception changes as the user moves towards or away from an object. This concept is neatly explained by Gibson, who writes "From an ecological point of view, the color of a surface is relative to the colors of adjacent surfaces; it is not an absolute color...For the natural environment is an aggregate of substances...the colors are not seen separately, as stimuli, but together as an arrangement" [14]. It is clear that human beings have developed a succinct *model* of the world, from thousands

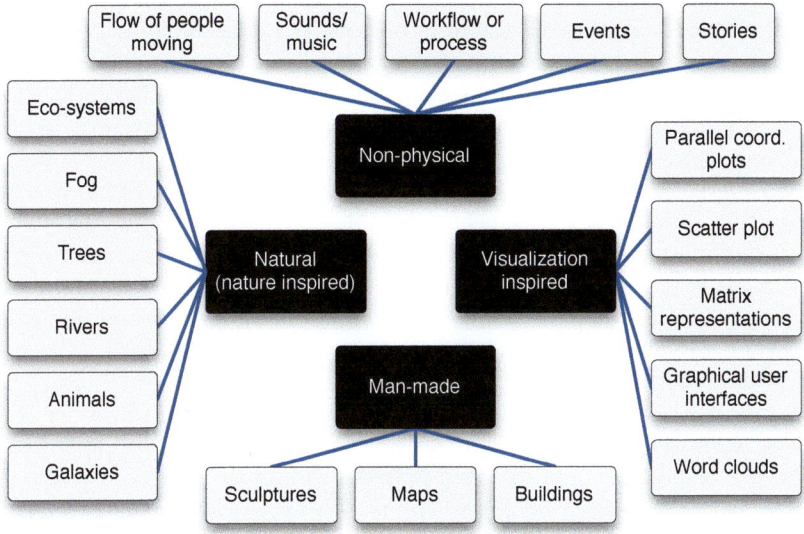

Fig. 7.1. We classify the metaphors into four types: Natural, non-physical, man-made and visualization inspired

of years of existence, and by being nurtured on Earth. This model can help visualization scientists develop succinct and effective visualizations. We can assume that a user will know the concepts (and indeed the *model*) of the natural world (say) and utilize this knowledge to immediately understand how we have transformed the data from an unseen version, into an appropriate visual mapping.

7.2 Classes of Metaphors

We group the metaphors into four categories: natural, non-physical, man-made and visualization inspired (see Fig. 7.1). While, these classifications provide a useful structure for this chapter, they can also be a mechanism for ideas. They can permit designers to 'think laterally' about different designs. For instance, a developer could ideate a visualization design 'inspired by a galaxy' and then provoke a different design by considering how the data could be displayed by a building.

7.2.1 Nature-Inspired

Many researchers have used objects from nature as inspiration for multivariate data and networks. The natural world contains very many concepts that could be used to ideate; these ideas range over massive scales from (say) the formation of galaxies and movement of tectonic plates to the growth of mold or the veins of a leaf. We use the term *natural* to represent things of the physical universe. For instance, we include geology and the Earth itself, wildlife that lives on our planet, and also some processes (such as weather) that affects the environments in which we live. Thus we include physical objects such as oil, sand, water, rocks, animals or rain, and also inanimate objects, such as wind, thunder, magnetism (as the Earth's magnetic field), gravity or weather. We exclude objects that have been substantially adapted by humans, such as plastics or glass, which we classify and include in the "man-made" category (Sect. 7.3), and also exclude non-physical processes that are not attributed to nature. For instance, the movement of humans through a building would be included in the "non-physical" category.

Natural metaphors offer convenient features that are often desirable to produce effective network visualizations. Nature is efficient. It utilizes the space economically. For instance, in a forest new trees spring up when old trees fall. They appear because light floods in from the canopy onto the forest floor, encouraging seeds that have fallen to sprout. Nature is also competitive. These new trees need to reach to the canopy roof as fast as they can. Nature is thus 'greedy'. The fittest and the fastest, or the strongest survive. In visualization we could imagine that the strongest cues or most representative data (to some metric) are pushed to the top of the visualization display, making them more noticeable. Nature is efficient. For instance, animals forage for food, they move in an efficient manner and cover the least amount of distance so to conserve energy. Nature utilizes the space efficiently. For instance, grasses grow to fill the space. Likewise, we can imagine that these positive effects can be used to make efficient desired space-filling techniques in visualization. Nature has therefore inspired several network visualizations, as discussed below.

Flowers provide a good source of inspiration for network visualization. Not only are they hierarchical, but also they are beautiful. For example, Chau [5] developed a flower metaphor for displaying web search results. The petals correspond to the number of key words, the leaves convey the number of outgoing links, the stem represents the length of the document, and the supporting ground indicates the number of incoming links. User studies showed that for complex data, the glyph outperformed a numeric based display. In fact, many other researchers have used the flower metaphor. Xiong and Donath [48] showed user interactions over time on a web messaging board, with time conveyed by the angle of the flower petal and color conveying the type of message. The stem length indicates how long a user has been at the message board. Zhu [53] used flower metaphors for both the people involved in communications as well as the discussion threads, while Van Loocke [43] used fractals to generate flowers and trees for multivariate binary data.

Grasses have also given inspiration. Although the examples of the Drift-Weed metaphor [33] were not on network data, this work provides a good example of using nature to inspire a visual metaphor for interactive analysis of multivariate data. The work enables different variables to be displayed along a line, with piecewise segments. The data is allocated onto the angles and lengths of the segments. Although individual grasses are difficult to understand, the overall effect appears as a single structure: a conglomeration of individual elements making one textured object. The overall texture object provides the user with an understanding of how the data trends.

As mentioned earlier, trees and plants have been used as inspiration for data visualization for many years. In 1991, Robertson, Mackinlay and Card presented 'Cone trees' [32]. Their seminal work visualized 10,000 files in the Unix file system using a vertical set of labels connected like an idealized three-dimensional tree. Their description of the system also included appropriate metaphoric language, writing "we provide operations for pruning and growing the view of a tree, collectively called gardening operations". In 2004, Shen and Eades [36] use a direct tree metaphor, in their MoneyTree system, where the trunk of the tree displaying trade volume and the leaves showing trade value, as shown in Fig. 7.2. Theron [42] used the rings of a tree to convey a hierarchical temporal data set. Data points are placed in rings based on their time stamp, and lines connect points that are part of the same branch of the hierarchy. While, recently Ma [25] used a tree as an egocentric depiction of a social network, where a 2D tree is divided left and right by the gender of the tree/person's contacts. The height of branches correspond to the age of the contact. Each leaf is a contact, and its position along the smallest branches indicates the date of the contact. Color and size of leaves can communicate additional information, as well as circular fruit along the small branches (see Fig. 7.3).

One of the more interesting dynamic behaviors one sees in nature is the flocking of birds, shoaling of fish, swarming of insects and herding with animals. Several researchers have used this notion to convey multivariate data and its relations. It is possible to utilize simple rules, such as separation, cohesion and alignment to create these behaviors. Vande Moere [27] used standard rules for flocking (collision avoidance, velocity matching, flock centering) and data relations (data similarity and dissimilarity) to render animations of stock market data over a period of time. Surfaces were wrapped around obvious clusters to both separate groups with different behavior and to expose outliers. Similarly, Yang et al. [51] use a flocking algorithm to examine networks of computers to help detect intrusions and other unusual behaviors. One view shows the topology of the network, and a second view flocks the network nodes based on a set of attributes associated with behavior and performance. Tekusova et al. [41] also focused on dynamics of stock market data. They combined flocking rules with data relations, much like Vande Moere. People flock as well, though in addition to the normal flocking rules, each individual may have a specific goal. Braun et al. [3] created a simulation of crowd motion that captures these two distinct types of goals (group, individual). They show how a crowd of virtual people can eventually exit a room with a single door. While these examples are not strictly used to inspire novel

Fig. 7.2. An example of MoneyTree, from Shen and Eades [36]. Top figure showing data from the 38th August 2003 at 10.50am, while the lower figure shows the stock market situation one hour later.

visualization designs, they do show that more elaborate rules can be used to drive the dynamics of the visualization.

Natural landscapes have inspired many visualization designs. Fabrikand, Montello and Mark [11] provide a useful critical analysis of different landscapes of the geographic domain, and we refer the reader to this work for a wider look at the use of landscape metaphors for information visualization in general. Designs that are positioned over a landscape provide an approachable metaphor for the user. Novice (and expert) users instantly understand

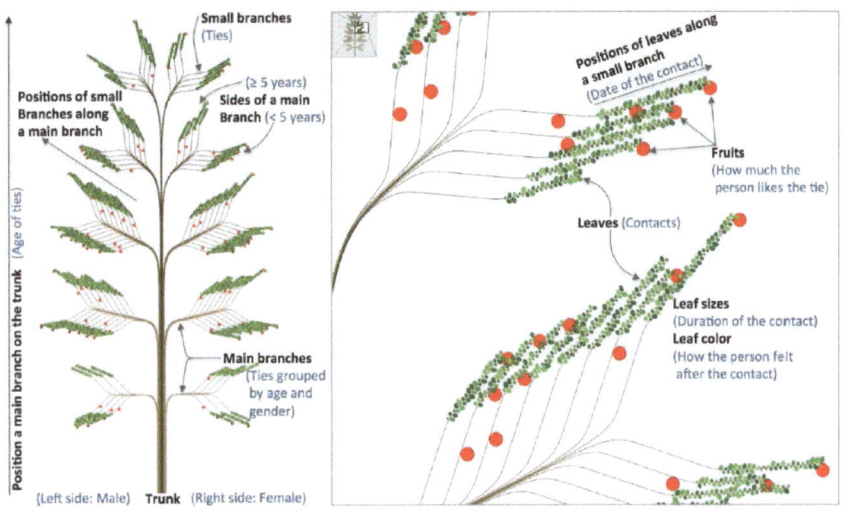

Fig. 7.3. An example of a ContactTree, showing the contacts of a single individual [25]

the metaphor and intuitively know how the information is presented in the space. They can start to make judgements over the position of data elements because position and nearness criteria are encoded in the information. In fact, many of the data presentations are naturally non-spatial. Developers are merely using this metaphor as a convenient way to display the data [38].

Wise [46], when referring to text visualization, names this an "ecological approach". There is a natural relationship between different parts (or organisms) to each other, and in the environment. In his work, the landscapes emerge because of the data, not because a designer has stipulated that the landscape will be of a particular form, but that the designer has encoded rules that determines the type of the visual depiction. For instance, the clusters with his Galaxies view, or the mountains of the ThemeScape^{TM} visualization [46] are created using self-organizing maps, which were trained on a sample set of documents. For network data specifically, Gansner et al. [12] display author collaboration data on a map layout. Xu et al. [49] present GraphScape, which they used to visualize multivariate network data of protein-protein interactions. Finally, the data from the development of large software systems has been displayed using the landscape metaphor. Balzer et al. say that "The familiar landscape metaphor facilitates intuitive navigation and comprehension [and their hierarchical approach enables] a clear representation of the relationships between the subsystems" [1].

It is clear that 'landscape' is a useful metaphor. In fact, secondary concepts such as water flow, or the natural flow and movement of something, has inspired developers. ThemeRiver represents an obvious example that follows a river metaphor. In their paper, Havre et al. [15] present a visualization technique that displays changes overtime, where larger effects are displayed in a wider bar. Symmetry is used to make the visualization more attractive. In their seminal paper, they use metaphoric language, describing the visualization as "directed flow from left to right", and "[the] width, of the river indicates a collective strength of the selected themes".

7.2.2 Non-physical

The idea of using rivers and flowing water leads us naturally to "non-physical" metaphors. We interpret the meaning of non-physical as something that does not contain physical matter; something that lacks substance and cannot be touched. In fact, many of the metaphors that we have already discussed, and the concepts that we are about to discuss in following chapters, have a physical space and are tangible in some way. They may be geographical (such as the nature inspired metaphors in Sect. 7.2.1), or created from man-made structures (in the next Sect. 7.3) or even visually tangible from the visualization-inspired section (Sect. 7.4). However there are many concepts that do not have a physical presence, which we consider in this section.

Let us consider our senses. The classic categorization is to describe five basic senses: vision, touch (including kinaesthetic and tactile), smell, taste, and sound. Obviously 'touch' is something physical, however let's also include vision as an artifact that is tangible (albeit on a computer screen). This leaves us with smell, taste and sound. Each of these senses encourage us to think emotionally. We could imagine therefore using these senses to inspire designs. For instance, consider being immersed in a soundscape of high quality audio sounds. The sounds could move around from the left to the right. This swaying could inspire a fluid visualization style where fluid is shown to flow around a network. We could imagine also ideas that were inspired from eating different foods. Carrots are crunchy, while mashed potatoes are smooth. We could imagine a visualization type that has smooth lines to represent one variable with angular parts mapped to another parameter. Before the reader concludes that the authors have lost sight of their objectives, we remind the reader of Eduard de Bono's [9] 'Lateral thinking', especially regarding provocation.

Other non-physical concepts that may be useful include: abstract processes, events, storytelling, motion of objects, and magnetism. In fact, storytelling is a good example of a useful non-physical metaphor. Stories are succinct forms of communication that are often interesting to hear, and therefore can be more memorable and easier to assimilate than other forms of communication. Gershon discusses what storytelling can do for visualization [13], suggesting that visualization is about finding effective visual metaphors; especially when presenting visualization results, stories can enable the developer to build the picture through a series of visualizations. Some visualizations are better to set the mood, place and time, while others build the story from disparate pieces of information to a final conclusion [13]. This style of visualization has been used to display different types of data. For instance, Walker et al. [45] use a comic-style storyboard to layout visualizations of microblog data. Segel and Heer [35] provide a systematic review of narrative visualization characterizing the designs with the narrative flow determined by the graphical interface and story discovery. Different stories appear from different users. In Chat Circles, Viegas and Donath [44] utilize simple metaphors or chatrooms to convey different social interactions from the online data.

7.3 Man-Made

It is easy to imagine that many man-made objects can be inspirational for network visualizations. Man-made objects contain many convenient and useful properties that can provide inspiration for visualization designs.

Let's start by looking at man-made structures and buildings. Designs for these objects benefit from properties of symmetry, regularity, repetition and recursion (self similarity). For instance, consider the design and construction of a modern house. These type of dwelling places are often built on a rectangular plan, with windows placed symmetrically: with windows on the ground-floor spaced directly below the windows on the first-floor. The brickwork is organized in regular and consistent patterns, and town housing or estates are constructed by a single contractor using only a few designs: the designs are used over and over again. Rooms within the buildings provide separation of content and often have specific roles (kitchen to cook, living area to relax, bedroom to sleep). 3D/Rooms by Card, Mackinlay and Robertson puts this idea into practice [4]. Each room holds different types of data. Users see an overview of all the rooms and choose where to go; they can locate objects of interest and walk through the rooms to another room.

Put many buildings and houses together and they form cities with many different types of buildings, some low and others very high. Earlier, in Sect. 7.2.1, we reflected on natural landscapes, but likewise man-made city landscapes and urban metaphors provide a useful structure to inspire network visualization designs. Such cityscapes have inspired several visualization designers. For example, Eick et al. [10] used the city metaphor to display software visualization of the quantity of changes indexed by developer in columns and module in rows, with a result that looks like a 3D bar-chart arrangement, while Keskin and Vogelmann use the cityscape metaphor to visualize hierarchical network data [23]. Finally the visualizations in the 'Selective Dynamic Manipulation of Visualizations' work by Chuah et al. [6] are shown as skyscrapers on a landscape.

Other man-made objects have caught the imagination of the visualization designer. Threads in fibers have inspired email visualization methods [21], and telescopes have inspired cylindrical visualization of hierarchical data [7].

Many of these visualizations are artistic. In fact, art, in general, and specifically artistic media such as oil paints and their brush strokes, have inspired many fluid flow visualization methods [24]. In addition to fluid flow visualization, various researchers have discussed the use of textures to represent multivariate data. For instance Interrante [16] discusses how different textures from paints can be used to depict multivariate data on a map. Different styles of illustration and drawings have been used in visualization, especially scientific visualization. For instance, Joshi et al. [18] use illustrative techniques inspired from art to display hurricane data, or caricaturistic-inspired visualization [30].

It is not only man-made physical objects, but also man-made non-physical processes can be inspirational. Pang uses a spray-can metaphor (spray, cut etc.) to interact with the visualization [29] while annealing in metallurgy (the process of heating and controlled cooling of the material to reduce artifacts) has inspired graph drawing algorithms to layout multivariate network data.

7.4 Visualization-Inspired

Finally, it is possible to conceive that visualization designs can be inspired by other visual designs. In sv3D [26], the authors write: "The sv3D Framework is a software visualization framework that builds on the SeeSoft metaphor". In this section we focus on three specific examples: glyphs, lists, and coordinated tag clouds and pixel oriented displays.

Brandes and Nick [2] work on exploring the evolution of dyadic (i.e., pairwise) relations in longitudinal social networks capturing asymmetric relations. Since dyadic evolution is the focus of the study, they propose gestaltlines, an intuitive visual metaphor to convey an entire dyadic evolution. The gestaltlines are gestalt-based use of glyphs in sparklines for multivariate sequences. They form a character-like visualization for each dyad and thus can be integrated into a matrix representation of all dyads in the network. The introduction of the glyph metaphor in this example allows depicting a longitudinal social network using a single, static image.

Jigsaw [40] is a visual analytics system aiming at helping analysts investigate the connections between entities across a report collection. The semantics of the entities are considered important, and an effective strategy toward the goal is to identify entities of interest and then conduct progressive exploration starting from them. A connected list metaphor is employed to achieve these requirements. In the List View of Jigsaw (see Fig. 7.4 for an example), the entities are organized into orderable lists by type. The links are displayed upon requests: after a user selects entities from the lists, all entities connected to them are highlighted and links are drawn between connected entities in adjacent lists. This intuitive view provides several functions that are critical for the goal: users can browse entities of a desired type, sorted either alphabetically or by frequency of appearance in different reports, to find interesting entities; they can investigate the immediate context of the interesting entities through interactive selections on the lists. It is difficult to

7.4 Visualization-Inspired

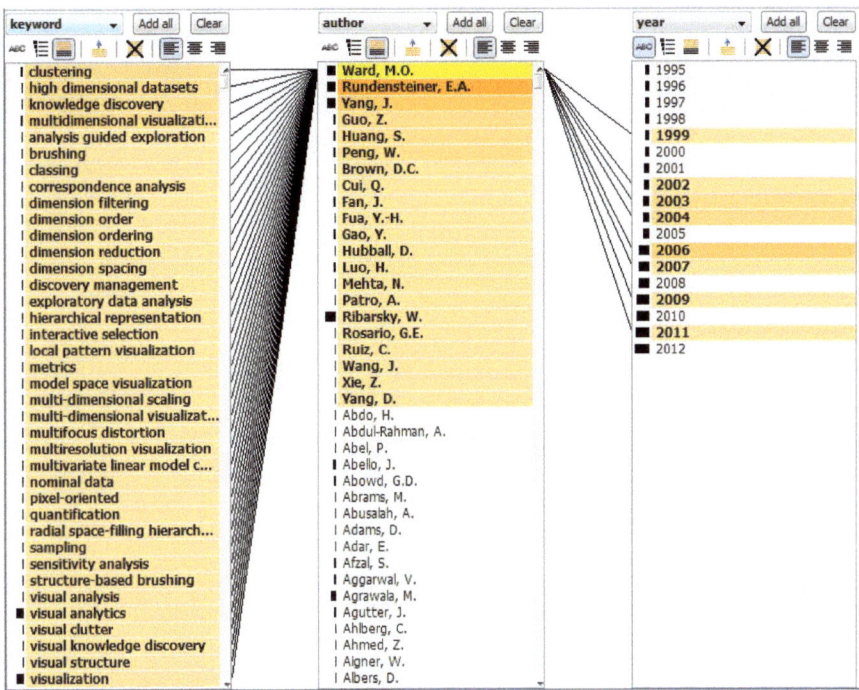

Fig. 7.4. An example of the List View of Jigsaw [40], showing the keywords, authors, and years relevant to Matthew Ward in a collection of InfoVis and VAST papers

conduct those tasks, especially the browsing task, on a node-link diagram, since the semantic contents will quickly clutter the display when the number of entities grows large.

Using tag clouds and pixel-oriented techniques, PIWI [50] helps analysts conduct Community-Related Tasks (CRT) for graphs with text labels. Examples of CRT include browsing the text labels of member nodes of many communities, exploring the relationships among multiple communities, analyzing the relationships between node attributes and the community structure, and selecting nodes according to attributes and the community structure. Since PIWI has a clear focus on community-related tasks, it picks communities as its information units. Two familiar visualization metaphors, namely tag clouds and pixel-oriented techniques [20], are then assembled to convey the semantics, structural, and attribute information of a graph (see Fig. 7.5 for an example). In particular, a community is represented by a tag cloud of text labels of its member nodes and a set of pixel-oriented node plots revealing the neighborhood information of the community. Thus, a graph is untangled into a set of communities that can be displayed row by row in a scrollable window.

Fig. 7.5. An example of PIWI [50]. The members and direct neighbors of a research community in a co-author network are highlighted in colorful tags. The tag cloud in a row displays the names of researchers in a community. The node plots besides the tag cloud depict the neighborhood information of the community and can be used for scalable selection.

In this way, a graph with thousands of labeled nodes becomes readable. The node attributes are also displayed using pixel-oriented node plots. By allowing users to conduct selections based on the node plots and integrate multiple selection results through Boolean operations, the users can conduct selection based on complex structural and attribute criteria. Since the users can interactively construct new communities and explore their contexts, progressive exploration is then supported. The novel metaphor of PIWI is inspired by the identification of communities as the essential information units according to the target tasks.

7.5 Proposed New Ideas

An effective metaphor is one in which a viewer can grasp instinctively how the information is being conveyed. Thus in considering new metaphors for multivariate graphs and trees we need to examine the human experience to find instances where the embedding of node and link/relation attributes requires little training. Below we outline a number of possibilities that, to the best of our knowledge, have not been used for this purpose. Clearly, many others exist; this is just a sampling based on the brainstorming of the authors.

7.5.1 Graphs/Networks

Many scenes we observe on a daily basis or have an intuitive feel for can be used to convey attributes of a multivariate network:

Students walking to class: For those of us in academia, a common scene is that of students moving along paths as they move from class to class. Each student can represent a node, and those nodes can have numerous attributes, such as gender, race, size, clothing color/style, hair color/style, and so on. They can move along paths (one link in, one link out) or stand at junctions (directional or non-directional links). The links themselves can convey relation attributes, using color, width, material type, and others to represent attribute values.

Cars on streets: Similar to students on paths, cars on streets is a regular phenomena that can be used for multivariate network visualization. Clearly the cars are nodes and the roads are links, with numerous attributes onto which data and relations can be mapped. To better convey the street metaphor, the dominant layout should be a grid, though as can be found in most cities, there are often roads that run diagonal to a grid or even form an overlay to the grid. Nodes and links that change their attributes are easily accommodated, though the addition/removal of nodes may result in a change in topology that might not be easy to incorporate. Removal of links would be comparable to closing a road due to construction.

Caves: While a cave requires the user to incrementally build an understanding of the entire graph/network, many games are based on the notion of exploring cave-like environments to discover connectivity and identify features of both nodes (rooms in the cave) and links (features of the tunnel). Node features could include the area or shape of a room, the height of the ceiling, or markings on the walls or floor. Link attributes could be the width or height of the tunnel, the texture of the floor or walls, the existence/frequency of lights, and so on. An interesting experiment would be to ascertain whether a graph learned via cave exploration is easier to remember than one shown as a traditional node-link diagram, given comparable time to explore it.

7.5.2 Hierarchies

Hierarchies/trees are also a common phenomena in what we see on a regular basis, and thus could be a vehicle for conveying a multivariate hierarchy. Some examples include:

Towns and buildings: Most cities and towns are laid out in hierarchies; a city contains neighborhoods or zip code areas, and these in turn consist of blocks bounded by streets. Buildings on these blocks can also be seen as hierarchies, as each building can have multiple floors, and each

Fig. 7.6. Some images of real scenes that might be usable to convey multivariate networks. All images courtesy of Wikimedia Commons.

floor can have multiple rooms. Rooms can contain many objects (e.g., furniture) that can be used to convey node information at the terminus of the hierarchy. Data for intermediate nodes and the links can be embedded in the hierarchy, using color, size, shape, and textures, for example. For deeper hierarchies, you can envision states containing multiple cities/towns, countries containing many states (different sizes, shapes, and positions), and countries forming nodes of a hierarchy rooted at the continent level. Clearly a single visualization could not adequately convey all these levels, but it would be relatively simple to provide navigation that would allow users to drill-down and roll-up for different levels of detail. Aggregation or summarization methods could be used to indicate the amount of information contained in the lower levels.

Clouds: Many types of cloud configuration can be viewed as a hierarchy, with smaller, nearer clouds being contained within larger, more distant clouds. While generally there is a fair amount of overlap between clouds at a comparable level (and of the same type), it would not be unheard of to have clouds at a given distance to be disjoint. Again, shape, color, size, texture, and opacity could all be used to embed information about nodes and relations. The resulting images may not look like real clouds, but would likely have sufficient similarity to reality to make viewing intuitive and interesting (we'd imagine that most people at some point in their lives have stared at clouds and wondered about their causes and relations).

7.6 Summary and Conclusions

In this chapter we described the use of metaphors as a mechanism to design and develop novel (non-node-link diagrams) methods for the visualization of multivariate networks. After defining what metaphors are and why they help us in developing new and often intuitive approaches to solving problems, we attempt to categorize different classes of metaphors that have been used to date to visualize graphs and hierarchies. We don't claim that this categorization is complete, but provides a good starting point for researchers in this area. We also suggest some ideas for metaphors that we don't believe have been used to date, yet are sufficiently familiar to people that we expect there would be only a modest learning curve for users to grasp the contents and relationships in visualizations generated via the metaphors.

The possibilities for new and intriguing metaphor-inspired interactive visualization system for network data is only limited by the imagination of the designer. The wealth of experiences we have been exposed to enables us to quickly draw analogies between our abstract data and phenomena we are exposed to on a regular basis, whether it be objects in our natural world, manufactured entities such as buildings and highway systems, non-physical objects and processes such as stories and magnetic forces, or even visualization methods not originally intended for network data. As multivariate networks are an abstraction on data values and relations, there are no preconceived notions of the best or most intuitive ways of projecting that information into a visual form. All that is needed is a creative mind and a willingness to think outside the box.

References

1. Balzer, M., Noack, A., Deussen, O., Lewerentz, C.: Software landscapes: visualizing the structure of large software systems. In: Proceedings of the Sixth Joint Eurographics/IEEE TCVG Conference on Visualization, VISSYM 2004, pp. 261–266. Eurographics Association, Aire-la-Ville (2004), http://dx.doi.org/10.2312/VisSym/VisSym04/261-266
2. Brandes, U., Nick, B.: Asymmetric relations in longitudinal social networks. IEEE Transactions on Visualization and Computer Graphics 17(12), 2283–2290 (2011)
3. Braun, A., Musse, S.R., de Oliveira, L.P.L., Bodmann, B.E.: Modeling individual behaviors in crowd simulation. In: 16th International Conference on Computer Animation and Social Agents, pp. 143–148. IEEE (2003)
4. Card, S.K., Robertson, G.G., Mackinlay, J.D.: The information visualizer, an information workspace. In: Proceedings of the SIGCHI Conference on Human Factors in Computing Systems, pp. 181–186. ACM (1991)
5. Chau, M.: Visualizing web search results using glyphs: Design and evaluation of a flower metaphor. ACM Transactions on Management Information Systems (TMIS) 2(1), 2 (2011)

6. Chuah, M.C., Roth, S.F., Mattis, J., Kolojejchick, J.: Sdm: Selective dynamic manipulation of visualizations. In: Proceedings of the 8th Annual ACM Symposium on User Interface and Software Technology, UIST 1995, pp. 61–70. ACM, New York (1995),
   http://0-doi.acm.org.unicat.bangor.ac.uk/10.1145/215585.215654
7. Dachselt, R., Ebert, J.: Collapsible cylindrical trees: A fast hierarchical navigation technique. In: IEEE Symposium on Information Visualization, INFOVIS 2001, pp. 79–86 (2001)
8. De Bono, E.: Six Thinking Hats. Penguin, UK (2009)
9. De Bono, E.: Lateral Thinking: Creativity Step by Step. HarperCollins (2010)
10. Eick, S.G., Graves, T.L., Karr, A.F., Mockus, A., Schuster, P.: Visualizing software changes. IEEE Transactions on Software Engineering 28(4), 396–412 (2002)
11. Fabrikant, S.I., Montello, D.R., Mark, D.M.: The natural landscape metaphor in information visualization: The role of commonsense geomorphology. J. Am. Soc. Inf. Sci. Technol. 61(2), 253–270 (2010),
    http://dx.doi.org/10.1002/asi.v61:2
12. Gansner, E.R., Hu, Y., Kobourov, S.G.: GMap: Drawing graphs as maps. In: Eppstein, D., Gansner, E.R. (eds.) GD 2009. LNCS, vol. 5849, pp. 405–407. Springer, Heidelberg (2010)
13. Gershon, N., Page, W.: What storytelling can do for information visualization. Commun. ACM 44(8), 31–37 (2001),
    http://doi.acm.org/10.1145/381641.381653
14. Gibson, J.J.: The Ecological approach to visual perception. Lawrence Erlbaum Associates (1979)
15. Havre, S., Hetzler, E., Whitney, P., Nowell, L.: Themeriver: visualizing thematic changes in large document collections. IEEE Transactions on Visualization and Computer Graphics 8(1), 9–20 (2002)
16. Interrante, V.: Harnessing natural textures for multivariate visualization. IEEE Comput. Graph. Appl. 20(6), 6–11 (2000),
    http://dx.doi.org/10.1109/MCG.2000.888001
17. Johnson, S.: Where good ideas come from: The natural history of innovation. Penguin, UK (2010)
18. Joshi, A., Caban, J., Rheingans, P., Sparling, L.: Case study on visualizing hurricanes using illustration-inspired techniques. IEEE Transactions on Visualization and Computer Graphics 15(5), 709–718 (2009),
    http://dx.doi.org/10.1109/TVCG.2008.105
19. Kang, H., Plaisant, C., Lee, B., Bederson, B.B.: Netlens: iterative exploration of content-actor network data. Information Visualization 6(1), 18–31 (2007)
20. Keim, D.A., Kriegel, H.P.: Visdb: A system for visualizing large databases. SIGMOD Rec. 24(2), 482 (1995), http://doi.acm.org/10.1145/568271.223895
21. Kerr, B.: Thread arcs: An email thread visualization. In: Proceedings of the Ninth Annual IEEE Conference on Information Visualization, INFOVIS 2003, pp. 211–218. IEEE Computer Society, Washington, DC (2003),
    http://dl.acm.org/citation.cfm?id=1947368.1947407
22. Kerren, A., Purchase, H.C., Ward, M.O.: Information Visualization – Towards Multivariate Network Visualization (Dagstuhl Seminar 13201). Dagstuhl Reports 3(5), 19–42 (2013), http://dblp.uni-trier.de/db/journals/dagstuhl-reports/dagstuhl-reports3.html#KerrenPW13

23. Keskin, C., Vogelmann, V.: Effective visualization of hierarchical graphs with the cityscape metaphor. In: Proceedings of the 1997 Workshop on New Paradigms in Information Visualization and Manipulation, NPIV 1997, pp. 52–57. ACM, New York (1997), http://0-doi.acm.org.unicat.bangor.ac.uk/10.1145/275519.275531
24. Laramee, R.S., Hauser, H., Doleisch, H., Vrolijk, B., Post, F.H., Weiskopf, D.: The state of the art in flow visualization: Dense and texture-based techniques. Computer Graphics Forum 23(2), 203–221 (2004), http://dx.doi.org/10.1111/j.1467-8659.2004.00753.x
25. Ma, K.L.: Visualization for studying social networks. In: Dagstuhl Seminar on Multivariate Network Visualization (2013)
26. Maletic, J.I., Marcus, A., Feng, L.: Source viewer 3d (sv3d): A framework for software visualization. In: Proceedings of the 25th International Conference on Software Engineering, ICSE 2003, pp. 812–813. IEEE Computer Society, Washington, DC (2003), http://dl.acm.org/citation.cfm?id=776816.776964
27. Moere, A.V.: Time-varying data visualization using information flocking boids. In: IEEE Symposium on Information Visualization, INFOVIS 2004, pp. 97–104. IEEE (2004)
28. Munzner, T.: A nested model for visualization design and validation. IEEE Transactions on Visualization and Computer Graphics 15(6), 921–928 (2009), http://dx.doi.org/10.1109/TVCG.2009.111
29. Pang, A., Clifton, M.: Metaphors for visualization. In: Sixth Eurographics Workshop on Visualization in Scientific Computing 1995, pp. 1–9. Springer (1995)
30. Rautek, P., Viola, I., Gröller, M.E.: Caricaturistic visualization. IEEE Transactions on Visualization and Computer Graphics 12(5), 1085–1092 (2006), http://dx.doi.org/10.1109/TVCG.2006.123
31. Roberts, J.C.: The Five Design-Sheet (FdS) approach for Sketching Information Visualization Designs. In: Maddock, S., Jorge, J. (eds.) Proc. Eurographics Education Papers, pp. 27–41. Eurographics Association (2011), http://diglib.eg.org/EG/DL/conf/EG2011/education/027-041.pdf
32. Robertson, G.G., Mackinlay, J.D., Card, S.K.: Cone trees: animated 3d visualizations of hierarchical information. In: Proceedings of the SIGCHI Conference on Human Factors in Computing Systems, CHI 1991, pp. 189–194. ACM, New York (1991), http://doi.acm.org/10.1145/108844.108883
33. Rose, S.J., Wong, P.C.: Driftweed: a visual metaphor for interactive analysis of multivariate data. In: Electronic Imaging, pp. 114–121. International Society for Optics and Photonics (2000)
34. Samet, H.: The design and analysis of spatial data structures, vol. 199. Addison-Wesley, Reading (1990)
35. Segel, E., Heer, J.: Narrative visualization: Telling stories with data. IEEE Transactions on Visualization and Computer Graphics 16(6), 1139–1148 (2010), http://dx.doi.org/10.1109/TVCG.2010.179
36. Shen, X., Eades, P.: Using moneytree to represent financial data. In: Proceedings of the Eighth International Conference on Information Visualisation (IV 2004), pp. 285–289 (2004)
37. Shneiderman, B.: Tree visualization with tree-maps: 2-d space-filling approach. ACM Trans. Graph. 11(1), 92–99 (1992), http://doi.acm.org/10.1145/102377.115768

38. Skupin, A., Fabrikant, S.I.: Spatialization methods: a cartographic research agenda for non-geographic information visualization. Cartography and Geographic Information Science 30(2), 95–119 (2003)
39. Stasko, J., Choo, J., Han, Y., Hu, M., Pileggi, H., Sadana, R., Stolper, C.D.: Citevis: Exploring conference paper citation data visually. In: Proceedings of the IEEE Conference on Information Visualization, Poster (2013)
40. Stasko, J., Görg, C., Liu, Z.: Jigsaw: supporting investigative analysis through interactive visualization. Information Visualization 7(2), 118–132 (2008)
41. Tekusova, T., Kohlhammer, J.: Applying animation to the visual analysis of financial time-dependent data. In: 11th International Conference on Information Visualization, IV 2007, pp. 101–108. IEEE (2007)
42. Therón, R.: Hierarchical-temporal data visualization using a tree-ring metaphor. In: Butz, A., Fisher, B., Krüger, A., Olivier, P. (eds.) SG 2006. LNCS, vol. 4073, pp. 70–81. Springer, Heidelberg (2006)
43. Van Loocke, P.R.: Generative flowers as a language of forms for the visualization of binary information. Leonardo 39(1), 9 (2006)
44. Viégas, F.B., Donath, J.S.: Chat circles. In: Proceedings of the SIGCHI Conference on Human Factors in Computing Systems, pp. 9–16. ACM (1999)
45. Walker, R., ap Cenydd, L., Pop, S., Miles, H.C., Hughes, C.J., Teahan, W.J., Roberts, J.C.: Storyboarding for visual analytics. Information Visualization (2013), http://ivi.sagepub.com/content/early/2013/05/27/1473871613487089.abstract
46. Wise, J.A.: The ecological approach to text visualization. J. Am. Soc. Inf. Sci. 50(13), 1224–1233 (1999), http://dx.doi.org/10.1002/(SICI)1097-4571(1999)50:13<1224::AID-ASI8>3.0.CO;2-4
47. Wiseman, R.: 59 seconds: think a little, change a lot. Random House Digital, Inc. (2010)
48. Xiong, R., Donath, J.: Peoplegarden: creating data portraits for users. In: Proceedings of the 12th Annual ACM Symposium on User Interface Software and Technology, pp. 37–44. ACM (1999)
49. Xu, K., Cunningham, A., Hong, S.H., Thomas, B.H.: Graphscape: integrated multivariate network visualization. In: 2007 6th International Asia-Pacific Symposium on Visualization, APVIS 2007, pp. 33–40. IEEE (2007)
50. Yang, J., Liu, Y., Zhang, X., Yuan, X., Zhao, Y., Barlowe, S., Liu, S.: Piwi: Visually exploring graphs based on their community structure. IEEE Trans. Vis. Comput. Graph. 19(6), 1034–1047 (2013)
51. Yang, L., Gasior, W., Katipally, R., Cui, X.: Alerts analysis and visualization in network-based intrusion detection systems. In: 2010 IEEE Second International Conference on Social Computing (SocialCom), pp. 785–790. IEEE (2010)
52. Young, J.W.: A Technique for Producing Ideas. Thinking Ink Media (2011)
53. Zhu, B., Chen, H.: Communication-garden system: Visualizing a computer-mediated communication process. Decision Support Systems 45(4), 778–794 (2008), http://www.sciencedirect.com/science/article/pii/S0167923608000195
54. Ziemkiewicz, C., Kosara, R.: The shaping of information by visual metaphors. IEEE Transactions on Visualization and Computer Graphics 14(6), 1269–1276 (2008)

8
Temporal Multivariate Networks

Daniel Archambault, James Abello, Jessie Kennedy, Stephen Kobourov, Kwan-Liu Ma, Silvia Miksch, Chris Muelder, and Alexandru C. Telea

In previous chapters, this book has primarily concerned itself with visualization methods for static, multivariate graphs. In a static scenario, the network has a number of attributes associated with its elements. These attribute values remain fixed and the challenge is to visualize the interactions between the network(s) and these attributes. Static multivariate graphs could be viewed as graphs with an associated high dimensional data set linked to its elements.

Time is simply another dimension in this multivariate data set that can interact with the vertices, edges, and attribute values of the network. However, humans perceive time differently as we know from our everyday interactions with the physical world. Thus, intuitively, this dimension is often handled differently when supporting the presentation of data that changes over time. Visualization applications and techniques have, and probably should, continue to exploit this fact, allowing for effective visualization methods of temporal multivariate graphs.

In this chapter, we define, characterize, and summarize the data and visualization techniques relating to temporal multivariate networks. Section 2.1.1 provides definitions and examples that characterize the networks we address in this chapter. We further refine our definitions of time in Sect. 8.2. In Sect. 8.3, we survey representations for dynamic multivariate networks and provide a survey of visualization techniques. We describe the visualization of temporal multivariate networks in the domain of software engineering in Sect. 8.4. Finally, Sect. 8.5 describes open problems in this area.

8.1 Definitions

In a variety of applications, time varying multivariate data can be viewed as evolving information networks whose structure is derived from data attributes (i.e., via similarity measures), is a specified a priory (i.e., the flow of information over an underlying network), or is the result of tracking behavioral statistics (i.e., network traces). The network and attributes can be:

- inherent to the fundamental data elements that are taken to be the network vertices *(name, age, gender, income, profession, interests, ...)*
- indicators of the type of relation between the network vertices *(professor of, father of, boss of, colleague of, ...)*
- attribute derived data *(time varying computational mappings from vertex attributes to edge attributes such as "pairs of stocks in markets whose performance has been above a given threshold during a time period")*
- structural derived statistics *(vertex ranks, network centrality, clustering measures, ...)*
- specified contexts in which the data occurs *(Tweets related to a given set of key words for a specified time period)*

In the next subsection, we adapt a model used in software engineering for the purposes of characterizing the types of dynamic, multivariate networks that can be visualized. Then, we propose mathematical formalizations of time varying multivariate networks.

8.1.1 Structure, Behavior, and Evolution

In a static multivariate network analysis scenario, we have a network structure, consisting of vertices and edges, as well as attributes associated with these vertices and edges. In a time varying scenario, both the graph structure and attribute values can evolve over time. In most cases, we can assume that the network structure at a given moment in time can *influence* how the attribute values evolve and vice versa. These interactions are in some respects very similar to those considered in some software engineering contexts [28]. Thus, we examine time varying multivariate networks appearing in biology and social networks under the lenses of *structure*, *behavior* and *evolution*.

- **Structure:** Pairings between elements of a complex system. Structure mostly relates to the topology of the underlying network at a given time t.
- **Behavior:** Observable activity. Action or reaction of system elements under a given set of stimuli. Behavior mostly refers to the *attributes* associated with the underlying network elements and how they change over time.
- **Evolution:** Gradual development of a configuration or pattern over time. Evolution mostly relates to the *structural* changes of the overall underlying network over time.

To illustrate these concepts, we provide examples in Table 8.1 drawn from the application areas considered in this book: biology, software engineering, and social networks. As an analogy to understand the overarching idea, consider a physical space, such as a building. The *structure* of the building is the construction at a given time. Its *behavior* is how people use the building and its rooms or interact with the physical structure. Its *evolution* may involve bringing in a construction crew to knock down walls and build new ones,

8.1 Definitions

Table 8.1. Examples of structure, behavior and evolution in the domains of biology, software engineering, and social networks

	Biology	Software Engineering	Social Networks
Structure	Biological entities, genes and interactions	Modules and couplings	A Twitter community network
Behavior	Gene expression levels	Program trace on the graph	Retweet, mention and follower activity
Evolution	Organism development; experimental conditions	Changes to the code	Changes to community structure

modifying the structure of the building as a result of observable decay in the physical infrastructure or as a response to ergonomical complaints of its occupants.

Note that in a time varying multivariate network scenario, both *behavior* and *evolution* can operate on each other. This generalization of the dynamics differs from the original software engineering definition where evolution could only influence behavior. An example of *evolution* influencing *behavior* in a biology scenario is when an experimental condition causes network structure to change or evolve, affecting in turn gene expression levels (i.e., behavior). An example of *behavior* influencing *evolution* in a social network scenario is when the interaction between actors in a social network (i.e., their behavior) causes ties to break or form, thus, evolving the network.

8.1.2 Formal Definitions of Temporal Multivariate Networks

To incorporate some of the main characteristics of time varying multivariate data we propose the following mathematical formalization of a time varying information data set [1].

The implicit assumption is that "time" is a universal reference "axis" with respect to which the data is being tracked. For now, we assume that "time" is a totally ordered set, but as we will discuss later, it can also be taken to be a partially ordered set. A *time varying information data set* $G_{V,t}$ on a set of vertices V consists of a sequence $\{F(G_t)\}_{t>=0}$ where F is a multivariate function $F : R^h \times R^h \times R \to R^k$ and at each time t, G_t denotes the following collection of 4-tuples:

$$G_t = \{< V_{(x,y)}, V_x, V_y, t >: V_{(x,y)} = F(V_x, V_y, t)\} \quad (8.1)$$

V_x, V_y, and $V_{(x,y)}$ are vectors in R^h and R^k respectively and (x,y) is a pair of vertices in V. The underlying *information network structure* is determined by those pairs of vertices (x,y) in $V \times V$ for which there exists a four tuple $< V_{(x,y)}, V_x, V_y, t >$ in some G_t.

The *F cumulative behavior* of $G_{V,t}$ up to and including t is the entry wise sum:

$$F_{<=t}(G_V) =< \sum_{j=0}^{t}(V_{(x,y)}), \sum_{j=0}^{t}(V_x), \sum_{j=0}^{t}(V_y) > \qquad (8.2)$$

where the sum is taken over all the quadruples $< V_{(x,y)}, V_x, V_y, j >$ in G_j for $j <= t$.

A time varying information data set $G_{V,t}$ *evolves towards a network* G, if there exists a time $t > 0$ such that the *underlying network* of the union of G_j for $j >= t$ is isomorphic to G.

8.2 Refining Our Models and Definitions for Time

Time itself is an inherent data dimension that is central to the tasks of revealing trends and identifying patterns and relationships in the data. Time and time-oriented data have distinct characteristics that make it worthwhile to treat such data as a separate data type [2, 3]. Due to the importance of time-oriented data, its structure has been studied in numerous scientific publications (e.g., [2, 11, 41]). As proposed by Aigner et al. [3], we divide the aspects of time-oriented data into general aspects required to adequately model the time domain as well as hierarchical organization of time and definition of concrete time elements, also called human-made abstractions.

The **general aspects** are scale, scope, arrangement, and viewpoints.

1. Scale: ordinal vs. discrete vs. continuous. As a first perspective, we look at time from the scale along which elements of the model are given. In an *ordinal* time domain, only relative order relations are present (e.g., before, after). In *discrete* domains temporal distances can also be considered. Time values can be mapped to a set of integers which enables quantitative modelling of time values (e.g., quantifiable temporal distances). Discrete time domains are based on a smallest possible unit and they are the most commonly used time model in information systems. *Continuous* time models are characterized by a possible mapping to real numbers, i.e., between any two points in time, another point in time exists (also known as dense time).

2. Scope: point-based vs. interval-based. Secondly, we consider the scope of the basic elements that constitute the structure of the time domain. *Point-based* time domains can be seen in analogy to discrete Euclidean points in space, i.e., having a temporal extent equal to zero. Thus, no information is given about the region between two points in time. In contrast to that, *interval-based* time domains relate to subsections of time having a temporal extent greater than zero. This aspect is also closely related to the notion of granularity, which will be discussed later.

3. Arrangement: linear vs. cyclic. As the third design aspect, we look at the arrangement of the time domain. Corresponding to our natural perception of time, we mostly consider time as proceeding *linearly* from the past to the future, i.e., each time value has a unique predecessor and successor. In a *cyclic* organization of time, the domain is composed of a set of recurring time values (e.g., the seasons of the year). Hence, any time value A is preceded and succeeded at the same time by any other time value B (e.g., winter comes before summer, but winter also succeeds summer).

4. Viewpoint: ordered vs. branching vs. multiple perspectives. The fourth subdivision is concerned with the views of time that are modelled. *Ordered* time domains consider things that happen one after the other. On a more detailed level, we might also distinguish between totally ordered and partially ordered domains. In a totally ordered domain only one thing can happen at a time. In contrast to this, simultaneous or overlapping events are allowed in partially ordered domains, i.e., multiple time primitives at a single point or overlapping in time. A more complex form of time domain organization is the so-called *branching* time. Here, multiple strands of time branch out and allow the description and comparison of alternative scenarios (e.g., in project planning). In contrast to branching time where only one path through time will actually happen, *multiple perspectives* facilitate simultaneous (even contrary) views of time.

The **human-made abstractions** are granularities, time primitives, and determinacy.

1. Granularity and calendars: none vs. single vs. multiple. To tackle the complexity of time and to provide different levels of granularity, useful abstractions can be employed. Basically, *granularities* can be thought of as (human-made) abstractions of time in order to make it easier to deal with time in every-day life (like minutes, hours, days, weeks, months). More generally, granularities describe mappings from time values to larger or smaller conceptual units. If a granularity and calendar system is supported by the time model, we characterize it as *multiple* granularities. Besides this complex variant, there might be a *single* granularity only (e.g., every time value is given in terms of milliseconds) or *none* of these abstractions are supported (e.g., abstract ticks).

2. Time primitives: instant vs. interval vs. span. These time primitives can be seen as an intermediary layer between data elements and the time domain. Basically, time primitives can be divided into anchored (absolute) and unanchored (relative) primitives. *Instant* and *interval* are primitives that belong to the first group, i.e., they are located on a fixed position along the time domain. In contrast to that, a *span* is a relative primitive, i.e., it has no absolute position in time. Instants are a model for single points in time, intervals for ranges between two points in time, and spans a duration (of intervals) without a fixed position.

3. Determinacy: determinate vs. indeterminate. Uncertainty is another important aspect when considering time-oriented data. If there is no complete or exact information about time specifications or if time primitives are converted from one granularity to another, uncertainties are introduced and have to be dealt with. Therefore, the *determinacy* of the given time specification needs to be considered. A determinate specification is present when there is complete knowledge of all temporal aspects.

8.3 Survey of Representations and Algorithms

While static graphs arise in many applications, dynamic processes naturally give rise to graphs that evolve through time. Such dynamic processes can be found in software engineering, telecommunications traffic, computational biology, and social networks, among others. Dynamic graph drawing addresses the problem of effectively presenting such relationships as they change over time.

Static graph visualization has a long and venerable history, while dynamic graph visualization is a relatively newer field. But even though temporal graph representations are more recent, the variety of representations is still large, and there are a number of studies concerning the drawing of dynamic graphs [5, 16, 20]. As a dynamic graph can be thought of as a sequence of edge sets on the same set of vertices, it can be treated similarly to visualizing multiple relationships on the same data set. There are nearly as many ways to represent dynamic or multivariate networks as there are graph representations: simple node-link diagrams, directed graphs, clustered graphs, hierarchical and multi-level representations, matrix representations, spatialized (map-like) representations, etc. Dynamic graphs can be visualized with *global views*, where all the graphs are displayed at once, *merged views*, where all the graphs are agglomerated together, and with *sequenced views*, where timesteps are plotted individually, and either small multiples or animated morphing (fading in/out vertices and edges that appear/disappear) are used to compare timesteps.

It is worth noting here that it makes a difference whether the temporal visualization aims to show individual timesteps (e.g., collaboration between researchers in each individual year) or cumulative (e.g., new collaborations from current year are added to the already accumulated collaboration graph). Similarly, there is a difference between offline and online temporal visualization. In the offline setting, we are given all data in advance, whereas in the online setting the changes are happening on the fly. Most existing algorithms address the problem of offline dynamic graph drawing, where the entire sequence of graphs to be drawn is known in advance. This gives the layout algorithm information about future changes in the graph, which makes it possible to optimize the layouts generated across the entire sequence (e.g., the algorithm can leave enough space in anticipation of placing vertices that appear later in the sequence). Less work has been done in the online setting, where the graph sequence to be laid out is not known in advance.

By far the most common method for visualizing dynamic graphs is to view the graph as a series of node-link diagrams whether as a sequence or all at once; see Fig. 8.1 and Fig. 8.2. Thus many dynamic graph layouts are based

8.3 Survey of Representations and Algorithms

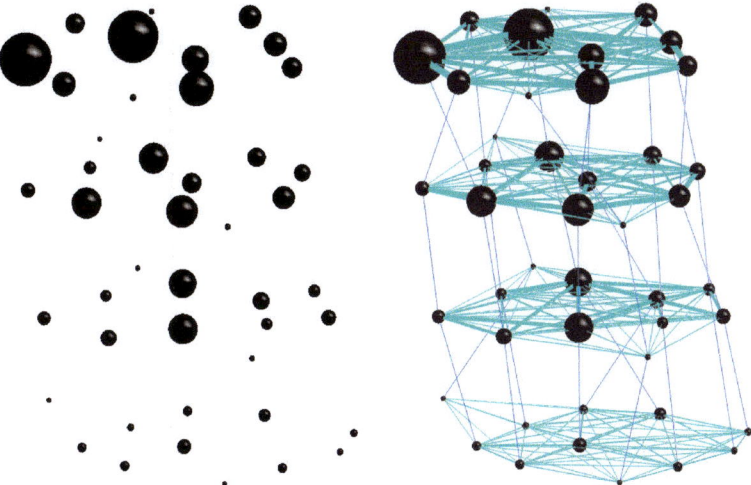

Fig. 8.1. A dynamic graph can be interpreted as a larger graph made of connecting graphs in adjacent timesteps [35]

on static graph layout algorithms, which are used to lay out each timestep. Efforts to improve the quality and stability of the layouts lead to the development of full-fledged dynamic graph layout algorithms. Some visualization approaches eschew the node-link representation to better show temporal evolution, as in streamline representations and dynamic maps. There has also been work in summarizing the temporal evolution of dynamic graphs in more static representations. And finally, there are a number of analytic algorithms and approaches that have been extended to dynamic network visualization.

8.3.1 Static Graph Layouts

Force-directed layouts (e.g., Fruchterman-Reingold [44], LinLog [78], Kamada-Kawai [63]) arrange graphs by iteratively refining the positions of vertices to incrementally reduce an energy function. This function varies between algorithms, but generally has the property that it is a function of the distances between vertices and the weights of the edges between them. These layouts are simple, and generally considered aesthetic, but they do not generally scale well to large or dense graphs.

More efficient layout algorithms use a multi-scale approach, such as the work of Cohen [23], the Fast Multipole Multilevel Method (FM^3) [52], and the Graph dRawing with Intelligent Placement (GRIP) algorithm [46]. These algorithms start by laying out a small approximation of a graph, then progressively laying out finer approximations of the graph, until the entire original graph is laid out. These algorithms generally use far fewer iterations, and

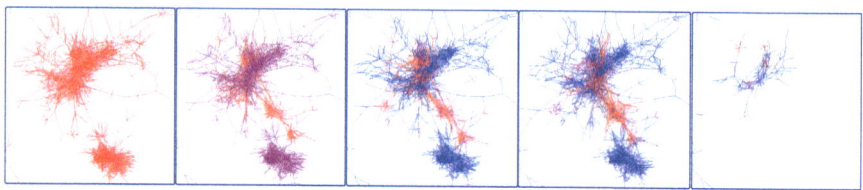

Fig. 8.2. Snapshots of the call-graph of a program as it evolves through time, extracted from CVS logs. Vertices start out red. As time passes and a vertex does not change it turns purple and finally blue. When another change is affected, the vertex again becomes red. Note the number of changes between the two large clusters and the break in the build on the last image [24].

thus perform far better than traditional force-directed approaches, while still producing similar results.

Even faster graph layout algorithms are available in the form of algebraic layouts, such as Algebraic Multigrid Computation of Eigenvectors (ACE) [65], High Dimensional Embedding (HDE) [54], the work of Brandes and Pich [18], or the Maxent method [48]. These calculate layouts directly using linear algebra techniques rather than using iterative force calculations. This generally makes them very fast. Clustering-based layouts have also been shown to be fast, as in the case of the treemap layout [75] or space-filling curve layout [74]. These methods work by clustering the graph in a preprocessing step and then mapping the clustering to the screen to define the layout itself.

8.3.2 Dynamic Graph Layouts

In dynamic graph drawing the goal is to maintain a nice layout of a graph that is modified via operations such as inserting/deleting edges and inserting/deleting vertices. A key property of in many real-world applications, where dynamic graphs naturally arise, is that the difference between any two timesteps is generally assumed to be incremental: that is, a small change relative to the size of the graph. If the change between timesteps is too large, then it is often more effective to treat them as separate, static networks. When visualizing evolving and dynamic graphs, two of the most important criteria to consider are:

1. *readability, or quality* of the individual layouts, which depends on aesthetic criteria such as display of symmetries, uniform edge lengths, and minimal number of crossings; and
2. *mental map preservation, or stability* in the series of layouts, which can be achieved by ensuring that vertices and edges that appear in consecutive graphs in the series, remain in the same location.

There is an inherent trade-off between the stability and quality of any dynamic graph layout, as restricting the movement of vertices could make it

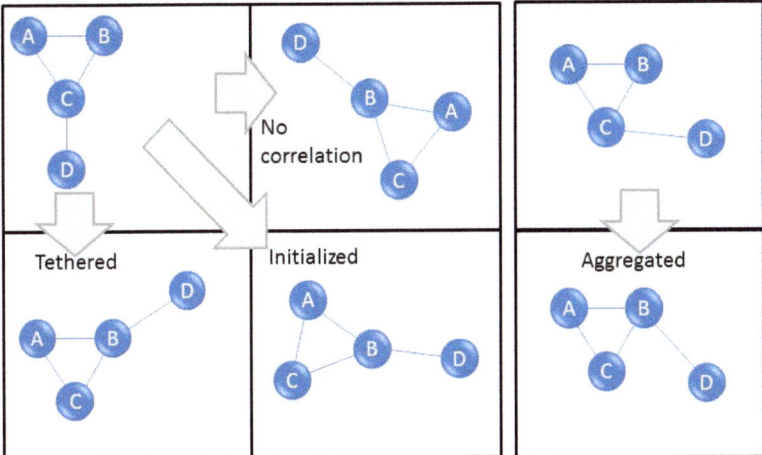

Fig. 8.3. Mental map preservation has been a forefront topic in dynamic graph layout. The level of layout stability can vary between approaches. Incremental approaches can range from having no correlation between timesteps to using the previous timestep as initialization to anchoring or tethering some vertices to previous positions. The most stable layouts agglomerate all timesteps together, but these could result in poor layouts at each timestep.

impossible to achieve high quality layout of the individual timesteps. In fact, these two criteria are often contradictory and many dynamic graph layout approaches explore different ways of balancing stability and quality; see Fig. 8.3. At one end are quality optimizing layouts with little to no correlation between timesteps, and at the other are fixed layouts where the vertices never move, even if the layout is not ideal for any given timestep. Anchored layouts lie somewhere between the two extremes, where some vertices are fixed while others are allowed to move; see survey of Brandes et al. [16].

The input to this problem is a series of graphs defined on the same underlying set of vertices. As a consequence, nearly all existing approaches to visualization of evolving and dynamic graphs are based on extensions of static graph layouts, usually based on a force-directed method. The simplest methods just initialize a force directed layout with the previous layout of the timestep, as in [10, 37], but this offers little guarantees for stability as nothing actually constrains the motion of vertices. Early examples of this can be dated back to North's DynaDAG [79], where the graph is not given all at once, but incrementally. Most of these early approaches, however, are limited to special classes of graphs and usually do not scale to graphs over a few hundred vertices. TGRIP could handle the larger graphs that appear in the real-world. It was developed as part of a system that keeps track of the evolution of software by extracting information about the program stored within a CVS version control system [24]. Such tools allow programmers to understand the evolution of a legacy program: Why is the program structured the way it is?

Which programmers were responsible for which parts of the program during which time periods? Which parts of the program appear unstable over long periods of time? TGRIP was used to visualize inheritance graphs, program call-graphs, and control-flow graphs, as they evolve over time; see Fig. 8.2.

Aggregate layouts such as in [71], are among the approaches that guarantee good stability by computing one layout for an aggregate graph made up of the union of all timesteps. Brandes and Corman [14] describe a system for visualizing network evolution in which both fixes vertices in constant locations, and uses a 3D super-graph, by showing each modification in a separate layer of a 3D representation with vertices common to two layers represented as columns connecting the layers. Thus, mental map preservation is achieved by pre-computing good locations for the vertices and fixing the position throughout the layers. An explicit tradeoff between quality and stability can also be provided as in the GraphAEL system [36]. There a supergraph of all timesteps is created and links between occurrences of the vertices in neighboring timesteps are added; see Fig. 8.1. By changing the weights of these inter-timestep edges one can emphasize stability (make inter-timestep edges very strong) or readability (make inter-timestep edges very weak). Such approaches [31, 34, 36, 39] generally use modified versions of traditional static layout algorithms directly, but often induce high memory usage and complexity because all timesteps are loaded at once. They are also only applicable to offline graph drawing, as the entire data range is needed at the beginning.

However, the most popular approach in recent years is to compute time varying network layouts by adding additional constraints that anchor vertices to their positions in the previous timestep [42, 43, 68]. These techniques work by adding some additional forces to the force direction calculation, but provide a good balance of cost, layout quality, and stability, and can be tuned by adjusting the anchor weights. These algorithms can also address the online dynamic graph drawing problem, as it is not necessary that the graph sequence is not known in advance. Brandes and Wagner adapt the force-directed model to dynamic graphs using a Bayesian framework [19]. An algorithm for visualizing dynamic social networks is discussed in [71]. Frishman and Tal consider dynamic drawing of clustered graphs [43] and of general graphs [42]. Brandes et al. have also performed a quantitative evaluation of the tradeoffs between layout quality and stability for these different classes of layouts [17].

There are also dynamic graph visualization approaches based on clustering. Kumar and Garland describe a method of animating clusters through time [66]. In this approach, a stratified, abstracted version of the graph is used, where the vertices are topologically sorted into a treelike structure (before layout) in order to expose interesting features.

Sallaberry et al. [95] cluster every timestep individually, associate the clusters across time, and use the space-filling curve approach to render each timestep; see Fig. 8.4. Pre-computing the clusters is computationally expensive. Hu et al. [59] propose a method based on a geographical metaphor to

8.3 Survey of Representations and Algorithms

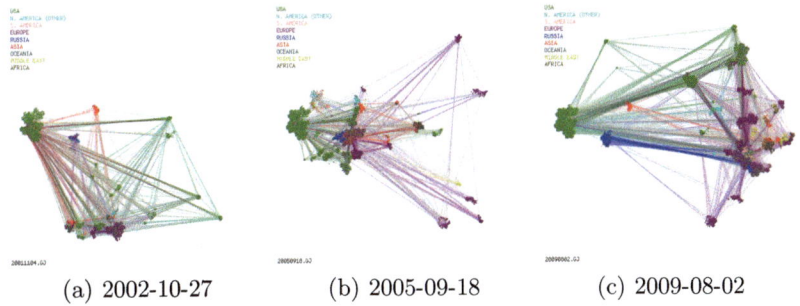

(a) 2002-10-27   (b) 2005-09-18   (c) 2009-08-02

Fig. 8.4. Large networks add additional challenges in computational cost and perceptual limits (images from [95])

visualize a summary of clustered dynamic graphs. It also relies on clustering and aims to keep clusters stable over time.

8.3.3 Animation versus Small Multiples

Often, dynamic graph visualizations animate the transitions between node-link diagrams of timesteps [13, 29, 36, 42, 49, 79]. In these animations, vertices dynamically appear, disappear and move to produce readable layouts at each timestep. Diehl and Görg [29] and Görg et al. [49] consider graphs in a sequence to create smoother transitions. Animations as a means to convey an evolving underlying graph have also been used in the context of software evolution [24] and scientific literature visualization [36]. Creating smooth animation between changing sequences of graphs is addressed using spectral graph visualization in [15]. When using the animation/morphing approach, it is possible to change the balance between readability of individual graphs and the overall mental map preservation, as in the system for Graph Animations with Evolving Layouts, GraphAEL [36, 40]. Applications of this framework include visualizing software evolution [24], social networks analysis [9], and the behavior of dynamically modifiable code [30].

Robertson *et al.* [90] evaluate the effectiveness of three trend visualization techniques. The results indicate that animation, often enjoyable and exciting, is not always well suited to data analysis. The other common alternative for visualizing multiple timesteps is to statically place them next to each other as small multiples [102]. This eases the comparison of distant timesteps but only a small area can be devoted to each timestep, which reduces the readability of each graph. Cerebral [8] is a system that uses a biologically guided graph layout and incorporates experimental data directly into the graph display. Small multiple views of different experimental conditions and a data-driven parallel coordinates view enable correlations between experimental conditions to be analyzed at the same time that the data is viewed in the graph context.

This combination of coordinated views allows the biologist to view the data from many different perspectives simultaneously.

Empirical studies to compare the advantages and drawbacks of these approaches ("Animation" vs. "Small Multiples") have been performed by Archambault et al. [7] as well as Farrugia and Quigley [38]. And even more recently, Rufiange et al. have developed a hybrid approach that lets the user interactively combine or switch between animations, small multiples, and plots that explicitly indicate what has changed [91].

8.3.4 Mental Map Preservation

Preserving the mental map, or layout stability, is a major focus in many dynamic node-link representations approaches [17, 42, 59, 66, 93]. Even though several experiments have been performed to examine the effect of preserving the mental map in dynamic graphs visualization the results are mixed. The results of [88] were quite surprising. The experiment found that the most effective visualizations were the extreme ones, i.e., the ones with very low or high mental map preservation, while visualizations with medium preservation were less effective [88]. With large networks, stability becomes even more important, but so does "motion coherency". Even small motions on each vertex are too much to perceive if they are chaotic, but if vertices move coherently, they can be perceived as a single group [95]. In a series of papers, Archambault and Purchase evaluate various approaches for dynamic graph visualization and consider how they affect mental map preservation [4, 6, 7], also summarized in a recent survey [5].

8.3.5 Alternative Representations

Using maps to visualize non-cartographic data has been considered in the context of spatialization [98]. Map-like visualization using layers and terrains to represent text document corpora dates back to 1995 [103]. The problem of effectively conveying change over time using a map-based visualization was studied by Harrower [55]. More recently, Mashima *et al.* [69] use the GMap framework [58] to visualize dynamic graphs with the geographic map metaphor; see Fig. 8.5.

Also related is work on visualizing subsets of a set of items. Areas of interest in a UML diagram can be highlighted using a deformed convex hull [22]. Isocontours-based bubblesets can be used to depict multiple relations defined on a set of items [25]. Automatic Euler diagrams, which show the grouping of subsets of items by drawing contiguous regions around them have also been considered [97]. Apart from differences in the algorithms used to generate regions, all of these approaches create regions that overlap with each other (unlike the strict map metaphor where regions do not overlap).

Bezerianos *et al.* [12] describe a multivariate network visualization system, GraphDice, which uses a plot matrix to navigate multivariate graphs.

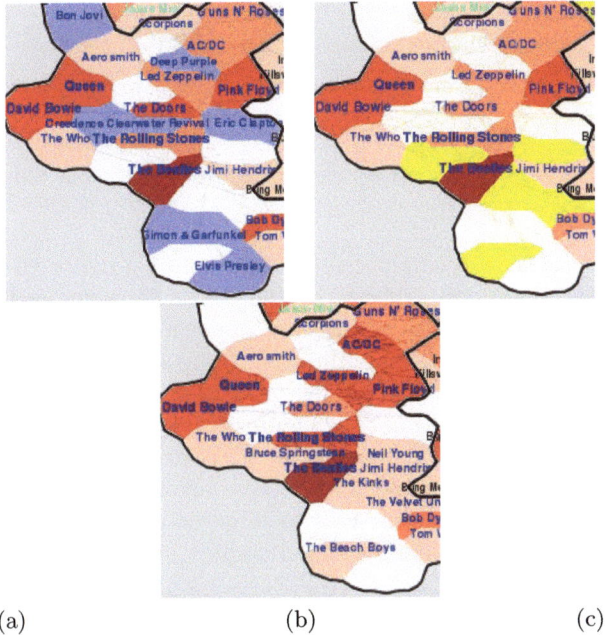

(a)   (b)   (c)

Fig. 8.5. Evolution in the top 250 most popular bands on Last.fm: showing three consecutive snapshots from an animation, focusing on area that corresponds to Rock. An animated version is also available online at http://www2.research.att.com/~yifanhu/TrendMap/. (a) Highlighting in blue areas where artists are about to disappear: Bon Jovi, Deep Purple, Elvis, Simon & Garfunkel, CCR, and Eric Clapton. (b) Highlighting in yellow the areas where new artists are about to appear. (c) An image after new artists appear, showing the newcomers: Bruce Springsteen, Neil Young, The Kinks, and The Beach Boys.

8.3.6 Static Temporal Plots

One visualization approach for summarizing dynamic large graphs is to directly represent time as an axis. The most direct way to do this is to take 2D node-link diagrams and extend them to 3D with time as the third dimension; see Fig. 8.1). However, 3D can be cluttered, and has occlusion and other perceptual limitations. An interesting 2D approach based on parallel coordinates was proposed by Burch et al. [21], where vertices are ordered and positioned on several vertical parallel lines, and directed edges connect these vertices from left to right. The graph of each timestep is thus displayed between two consecutive vertical axes.

Such representations can get quite cluttered for larger graphs. Rather than depicting the entire network over time, another approach is to abstract the network into clusters and to show how they evolve. WilmaScope [32] does this in 3D by representing the clusters as tubes. An increasingly popular way to

Fig. 8.6. Storylines can succinctly summarize the evolution of a dynamic graph (from [99])

visualize the evolution dynamic clusters is the use of storylines [27, 64, 77, 80, 89, 99]. Most of these works reference hand-drawn diagrams such as XKCD's movie narrative charts [77] as inspiration, in which entities are represented as lines which move together when in the same group and separate when they are not. Plotweaver [81] is a tool to aid in semi-automatic generation of storyline plots, but it still requires significant user interaction. The works of Ogawa et al. [80] and Tanahashi et al. [99] aim to automate the process; see Fig. 8.6. However, producing good results with these algorithms is computationally expensive, as they do not scale well to large data sets. To apply storyline techniques to dynamic graphs, an intermediary step of dynamic clustering must be derived [89, 95].

8.3.7 Dynamic Graph Analytics

Another relevant avenue of research has been the extension of analytic algorithms to dynamic graphs. Finding a partition of the vertices of a static graph according to its structure is a well studied problem; see survey by Schaeffer [96]. But clustering a dynamic graph is a less studied problem. One possibility is to use a global clustering, which is computed by applying a static clustering to an aggregate combination of all the timesteps in the dynamic graph. This creates a clustering which is on average good, but which can not capture the evolution of the network. Others have developed dynamic graph clustering algorithms in the context of visualization applications that track clusters across timesteps, allowing their memberships to evolve over time. Several approaches try to modify the clustering incrementally as the network changes [50, 51, 94]. Hu et al. [59] use a similar approach, but apply a heuristic to accelerate this process. Sallaberry et al. [95], on the other hand, cluster each timestep separately and then use Jaccard index to track the clusters across time.

Different from top-down methods above, there are also several bottom-up approaches that start with a single vertex and its immediate context. Additional relevant vertices and connections are revealed only on demand, based on graph structure or specialized degree-of-interest functions that can incorporate semantic importance or users' interaction histories [26, 33, 45, 53, 56, 73]. Recently, such approaches have been extended to dynamic graphs by incorporating temporal histories, and applying relevancy filtering to a storyline-based representation [76].

8.4 Applications to Software Engineering

Temporal multivariate networks play a key role in many aspects of software engineering (SE). To understand the related challenges, we need to understand

1. the *tasks* that they support in software engineering;
2. the *characteristics* of SE data leading to such graphs.

This section covers the above two points. For a full overview of applications of multivariate dynamic graphs in SE, we refer to Chap. 2. Our focus here is more technical. Specifically, we aim to characterize SE graphs from the perspective of time modeling (Sect. 8.2), and the variability axes (of types) (Sect. 8.3). This in turn better explains the rationale behind the visual designs presented in Chap. 2, and also why it is challenging to use visualization techniques developed for other types of temporal multivariate graphs to handle SE graphs.

Tasks

Software engineering activities cover the entire software product lifetime, starting with requirement gathering, followed by architecting, design, implementation (coding), testing, release, and ending with maintenance. Graphs are created and used in all these stages, as shown in Table 8.2. As software systems change during their lifecycle, all above graphs are by nature time-dependent. Moreover, SE graphs involve elements and relations spanning several of the above activities. For example, in reverse engineering, we encounter graphs that link software test results with source code (and developers), class diagrams, and requirements.

Table 8.2. Examples of multivariate temporal graphs in SE

Actions	Examples of graphs
Requirements	Requirements vs. tasks vs. stakeholders [61]
	UML use-case diagrams [92]
Architecting	System structure (layering, dataflows, component interactions) [101]
Design	UML component and package diagrams [92]
	UML class, activity diagrams [92]
Coding	Call, inheritance, type-use, and include graphs [28]
Testing	Type-instance graphs, control flow graphs [86]
	Resource allocation graphs [72]
Release	Deployment graphs [82], UML deployment diagrams [92]
Maintenance	Developer networks, code duplication graphs [70]

Data Characteristics

Temporal multivariate SE graphs have several characteristics which make their computation, efficient manipulation, and above all *understanding* very challenging. Below we outline the main such aspects.

Size: Depending on their type, SE graphs range from a few tens of elements (UML diagrams and developer networks) to hundreds of thousands (call graphs) or even millions of elements (control-flow graphs of large programs). The static call graph of the Mozilla Firefox browser (a medium-sized system as compared to large telecom or banking software) has, for example, over 500K edges [57]. Certain topology constraints exist for some graphs, e.g. class hierarchies are, usually, trees, and architecture dependencies form a directed acyclic graph. However, in the general case, little can be said about the global properties of SE graphs. For instance, a call graph can be cyclic (or not), and can have a widely varying distribution of number of edges per vertex depending on application type.

Attributes: Each vertex and edge in a SE graph typically has several attributes. These describe both static and dynamic properties of the entity encoded by that vertex or edge. For instance, annotated semantic graphs (ASGs) for C++ programs have tens of such attributes [100]. Computing software quality metrics easily adds tens of other metrics [67]. Attribute types span a wide spectrum: numerical, categorical, text, and binary. Attribute *types* are key to effective program understanding. For instance, the C++ ASG in [100] contains around two hundred different vertex-attribute types that encode the different properties of the annotated C++ grammar. Being able to visually distinguish between different types is essential, e.g., for detecting the presence of specific design or execution patterns. Missing values are possible, e.g., due to limitations of program analysis tools or due to incomplete program coverage for execution monitoring tools.

Dynamics: Graphs describing human aspects, such as developer activity, change slowly, given the continuous nature of software evolution [70]. However, other SE graphs exhibit different dynamics. For instance, in program execution graphs, large changes can occur in short time periods and few changes in other longer time periods. Dynamics is present both at the *structure* level (e.g., changes of a call graph topology as the program is run for different inputs or as code changes during maintenance), and also at the *attribute* level (e.g., different runtime metrics measured at static component level for different program executions).

Time Modeling: Time is, formally, modeled as a discrete quantity, since both execution and changes of software code occur at discrete, moments. Time has a linear nature, describing the order of execution of program instructions or the order of changes in a repository. However, time can be seen as fully ordered or branching (Sect. 8.1). The branching case occurs, e.g., when considering execution of multi-threaded programs or analyzing development activity of a repository with multiple branches. Both point-based and interval-based models are used, often interchangeaby, for the same analysis. For instance, a *version* in a software repository can refer to the moment when it was committed, but also to the time interval between this commit time and the next change of the same artifact.

Scale: Software understanding occurs on multiple levels of detail and following both a top-down and bottom-up process [86]. Hence, one needs to (visually) analyze software at several levels of detail or *scales*. SE graphs offer several natural scales, given their hierarchical, or compound, nature (Chap. 2), e.g., function-class-file-folder or the structure given by a function call stack. Yet, several aspects make constructing efficient and effective multiscale SE-graph visualizations hard. Firstly, SE graphs are huge. The few above-mentioned levels of detail do not offer enough granularity to *automatically* simplify large graphs to levels where they can be displayed in an understandable manner. Automatically computing *additional* levels of detail is hard – for instance, what should be the meaning of an artifact larger than a file, but smaller than a folder? Secondly, many program understanding tasks require showing both fine-grained detail *and* coarse-scale structure in the same view. For instance, to debug a crash, we need to see the entire call stack, from the finest-grained instruction which caused the fault up to the coarse-level components which scope the fault. Finally, software is by nature *abstract*. As such, finding effective visual metaphors (for both the spatial graph embedding and attribute mapping) is challenging.

8.5 Open Problems

Although significant progress has been achieved in the design of visualization methods and tools for exploring multivariate temporal networks, several important open challenges remain. This section outlines a selection of challenges which are relevant to a broad subset of applications involving such graphs. Throughout the discussion, we use the notation introduced in Sect. 8.1.2.

8.5.1 Attribute Dimensionality

As outlined in Sect. 8.4, SE graphs are high-variate, i.e., have many attributes for each vertex or edge. Existing visualization techniques can simultaneously show a few (up to three) attributes per graph element, by mapping these to shape, size, texture, color, and shading. However, this solution scales poorly for graphs of hundreds of thousands of elements. Separately, even for small graphs (hundreds of elements), showing tens of attributes per element is an open challenge. Parallel coordinate plots partially address this quest [60]. An interesting adaption hereof clusters graph vertices based on attribute values, and links the resulting icicle plots to a table-lens-like visualization of the edge attributes, to highlight attribute correlations [87]. Dimensionality reduction projects a set of high-dimensional attributes into $\mathbb{R}^2$ or $\mathbb{R}^3$ so that similarities between the original attributes are reflected in the low-dimensional distance [62, 83, 84]. Although such approaches scale well computationally for large sample counts [85], it is hard to visualize both attribute similarity *and* graph structure in the same embedding. Other approaches use interactive brushing, attribute selection, and linked views. However, none of the above methods fully enables users to correlate structure with attributes, and attributes among themselves, for highly-variate graphs.

8.5.2 Capturing Patterns

In many use-cases, showing a picture of the (changing) graph is not sufficient, even when this picture is clutter and overlap-free. For instance, consider the task of locating *patterns* in the graph. Patterns are specific configurations of vertices and edges (topology) and attribute values which capture events of interest. Patterns are typically problem-dependent, and have a certain variability in both structure and attribute values. Consider finding a 'multithreading refactoring event' in a software code base: This would involve finding similar code fragments in a graph G_t, which describe serial code, and finding that they have been replaced by functionally-identical multithreaded code in the following revision G_{t+1} of the code base. Even the simpler 'design patterns' [47], well known and used in object-oriented software design, are hard to detect and visualize. The underlying reasons are twofold. First, patterns involve, by definition, *several* vertices, edges, and attribute values, so they correspond to portions of a graph visualizations. However, existing graph visualization techniques have difficulties in showing such data subsets in canonical ways, i.e., in ways that make their visual detection easy. Secondly, patterns have a certain variability. Besides making automatic detection hard, this also implies that their graph visualizations will exhibit a necessary variability, which makes their visual detection hard. Finally, visually detecting dynamic patterns is very challenging—if animation is used, this poses high demands on the user's visual memory; if static visualizations are used, inherently dynamic patterns may be hard to grasp.

8.5.3 Data Size

Large dynamic graphs involve large sets of vertices and edges and/or many sampling moments when the graph is captured. This implies many sample points taken over the domain of function F (Eq. 8.1). Large graphs are hard to embed in a low-dimensional space ($\mathbb{R}^2$ or $\mathbb{R}^3$) so that the graph structure is easy to discern. This basic graph-drawing problem becomes one or two orders of magnitude larger for dynamic graphs. The data size problem becomes even larger for high-variate graphs.

It is insightful to consider how data size relates to the other challenges. Formally, we could argue that dynamic multivariate graphs (and their patterns) can be efficiently and effectively depicted using existing visualization methods, for small graphs. Hence, we could use subsampling, like in scientific data visualization, to reduce the graph size prior to visual exploration. To preserve features or patterns of interest, data-adaptive subsampling could be used. The main obstacle here is that we still lack a comprehensive theory for subsampling graphs *and* categorical attributes. As such, existing solutions addressing data size currently have to rely on aggregation and simplification algorithms and heuristics that are problem, scale, and even dataset-specific.

8.6 Summary and Conclusions

In this chapter, we characterized temporal multivariate graphs in terms of structure and time. We presented common terminology for discussing temporal multivariate graphs, a survey of existing techniques, focusing on software engineering applications, and a collection of open problems. We hope that this common terminology, data characterization, and organization of existing and future work will help foster further research in the emerging area of dynamic multivariate graph visualization.

References

1. Abello, J., Hadlak, S., Schumann, H., Schulz, H.: A modular degree-of-interest specification for the visual analysis of large dynamic networks. IEEE Transactions on Visualization and Computer Graphics (in press, 2014)
2. Aigner, W., Miksch, S., Schumann, H., Tominski, C.: Visualization of Time-Oriented Data. Springer, London (2011)
3. Andrienko, N., Andrienko, G.: Exploratory Analysis of Spatial and Temporal Data: A Systematic Approach. Springer, Berlin (2006)
4. Archambault, D., Purchase, H.C.: The mental map and memorability in dynamic graphs. In: Hauser, H., Kobourov, S.G., Qu, H. (eds.) Proc. of the IEEE Pacific Visualization Symposium, pp. 89–96. IEEE (2012)
5. Archambault, D., Purchase, H.C.: The "map" in the mental map: Experimental results in dynamic graph drawing. International Journal of Human-Computer Studies 71(11), 1044–1055 (2013)
6. Archambault, D., Purchase, H.C.: Mental map preservation helps user orientation in dynamic graphs. In: Didimo, W., Patrignani, M. (eds.) GD 2012. LNCS, vol. 7704, pp. 475–486. Springer, Heidelberg (2013)
7. Archambault, D., Purchase, H.C., Pinaud, B.: Animation, small multiples, and the effect of mental map preservation in dynamic graphs. IEEE Transactions on Visualization and Computer Graphics 17(4), 539–552 (2011)
8. Barsky, A., Munzner, T., Gardy, J., Kincaid, R.: Cerebral: Visualizing multiple experimental conditions on a graph with biological context. IEEE Transactions on Visualization and Computer Graphics 14(6), 1253–1260 (2008)

9. Bastian, M., Heymann, S., Jacomy, M.: Gephi: an open source software for exploring and manipulating networks. In: International AAAI Conference on Weblogs and Social Media, pp. 361–362 (2009)
10. Bender-deMoll, S., McFarland, D.A.: The art and science of dynamic network visualization. Journal of Social Structure 7(2) (2006)
11. Bettini, C., Jajodia, S., Wang, S.X.: Time Granularities in Databases, Data Mining, and Temporal Reasoning. Springer, Berlin (2000)
12. Bezerianos, A., Chevalier, F., Dragicevic, P., Elmqvist, N., Fekete, J.D.: Graphdice: A system for exploring multivariate social networks. Computer Graphics Forum 29(3), 863–872 (2010)
13. Boitmanis, K., Brandes, U., Pich, C.: Visualizing internet evolution on the autonomous systems level. In: Hong, S.-H., Nishizeki, T., Quan, W. (eds.) GD 2007. LNCS, vol. 4875, pp. 365–376. Springer, Heidelberg (2008)
14. Brandes, U., Corman, S.R.: Visual unrolling of network evolution and the analysis of dynamic discourse. In: Proc. of the IEEE Symposium on Information Visualization, pp. 145–151 (2002)
15. Brandes, U., Fleischer, D., Puppe, T.: Dynamic spectral layout with an application to small worlds. Journal of Graph Algorithms and Applications 11(2), 325–343 (2007)
16. Brandes, U., Indlekofer, N., Mader, M.: Visualization methods for longitudinal social networks and stochastic actor-oriented modeling. Social Networks, 291–308 (June 2011)
17. Brandes, U., Mader, M.: A quantitative comparison of stress-minimization approaches for offline dynamic graph drawing. In: Speckmann, B. (ed.) GD 2011. LNCS, vol. 7034, pp. 99–110. Springer, Heidelberg (2011)
18. Brandes, U., Pich, C.: An experimental study on distance-based graph drawing. In: Tollis, I.G., Patrignani, M. (eds.) GD 2008. LNCS, vol. 5417, pp. 218–229. Springer, Heidelberg (2009)
19. Brandes, U., Wagner, D.: A Bayesian paradigm for dynamic graph layout. In: DiBattista, G. (ed.) GD 1997. LNCS, vol. 1353, pp. 236–247. Springer, Heidelberg (1997)
20. Branke, J.: Dynamic graph drawing. In: Kaufmann, M., Wagner, D. (eds.) Drawing Graphs. LNCS, vol. 2025, pp. 228–246. Springer, Heidelberg (2001)
21. Burch, M., Vehlow, C., Beck, F., Diehl, S., Weiskopf, D.: Parallel edge splatting for scalable dynamic graph visualization. IEEE Transactions on Visualization and Computer Graphics 17(12), 2344–2353 (2011)
22. Byelas, H., Telea, A.: Visualization of areas of interest in software architecture diagrams. In: ACM SoftVis 2006, pp. 105–114 (2006)
23. Cohen, J.D.: Drawing graphs to convey proximity: An incremental arrangement method. ACM Transactions on Computer-Human Interaction 4(3), 197–229 (1997)
24. Collberg, C., Kobourov, S.G., Nagra, J., Pitts, J., Wampler, K.: A system for graph-based visualization of the evolution of software. In: ACM SoftVis 2003, pp. 77–86 (2003)
25. Collins, C., Penn, G., Carpendale, S.: Bubble sets: Revealing set relations with isocontours over existing visualizations. IEEE Transactions on Visualization and Computer Graphics 15(6), 1009–1016 (2009)
26. Crnovrsanin, T., Liao, I., Wuy, Y., Ma, K.L.: Visual recommendations for network navigation. In: Proc. of the 13th Eurographics/IEEE - VGTC Conference on Visualization, EuroVis 2011, pp. 1081–1090. Eurographics Association, Aire-la-Ville (2011)

27. Cui, W., Liu, S., Tan, L., Shi, C., Song, Y., Gao, Z., Qu, H., Tong, X.: Textflow: Towards better understanding of evolving topics in text. IEEE Transactions on Visualization and Computer Graphics 17(12), 2412–2421 (2011)
28. Diehl, S.: Software Visualization: Visualizing the Structure, Behaviour, and Evolution of Software. Springer, Berlin (2010)
29. Diehl, S., Görg, C.: Graphs, they are changing. In: Goodrich, M.T., Kobourov, S.G. (eds.) GD 2002. LNCS, vol. 2528, pp. 23–30. Springer, Heidelberg (2002)
30. Dux, B., Iyer, A., Debray, S.K., Forrester, D., Kobourov, S.G.: Visualizing the behavior of dynamically modifiable code. In: IWPC, pp. 337–340 (2005)
31. Dwyer, T., Gallagher, D.R.: Visualising changes in fund manager holdings in two and a half-dimensions. Information Visualization 3, 227–244 (2004)
32. Dwyer, T.: Extending the WilmaScope 3D Graph Visualisation System – Software Demonstration. In: Hong, S.H. (ed.) APVIS. CRPIT, vol. 45, pp. 39–45. Australian Computer Society (2005)
33. Elmqvist, N., Fekete, J.D.: Hierarchical Aggregation for Information Visualization: Overview, Techniques, and Design Guidelines. IEEE Transactions on Visualization and Computer Graphics 16(3), 439–454 (2009)
34. Erten, C., Kobourov, S., Le, V., Navabi, A.: Simultaneous graph drawing: layout algorithms and visualization schemes. Journal of Graph Algorithms and Applications 9(1), 165–182 (2005)
35. Erten, C., Harding, P.J., Kobourov, S.G., Wampler, K., Yee, G.: Exploring the computing literature using temporal graph visualization. In: Electronic Imaging 2004, pp. 45–56 (2004)
36. Erten, C., Harding, P.J., Kobourov, S.G., Wampler, K., Yee, G.: GraphAEL: Graph animations with evolving layouts. In: Liotta, G. (ed.) GD 2003. LNCS, vol. 2912, pp. 98–110. Springer, Heidelberg (2004)
37. Farrugia, M., Quigley, A.: Cell phone mini challenge: Node-link animation award animating multivariate dynamic social networks. In: IEEE Visual Analytics Science and Technology, pp. 215–216 (October 2008)
38. Farrugia, M., Quigley, A.: Effective temporal graph layout: A comparative study of animation versus static display methods. Journal of Information Visualization 10(1), 47–64 (2011)
39. Feng, K.C., Wang, C., Shen, H.W., Lee, T.Y.: Coherent time-varying graph drawing with multi-focus+context interaction. IEEE Transactions on Visualization and Computer Graphics (2011)
40. Forrester, D., Kobourov, S.G., Navabi, A., Wampler, K., Yee, G.V.: Graphael: A system for generalized force-directed layouts. In: Pach, J. (ed.) GD 2004. LNCS, vol. 3383, pp. 454–464. Springer, Heidelberg (2005)
41. Frank, A.U.: Different Types of "Times" in GIS. In: Egenhofer, M.J., Golledge, R.G. (eds.) Spatial and Temporal Reasoning in Geographic Information Systems, pp. 40–62. Oxford University Press, New York (1998)
42. Frishman, Y., Tal, A.: Online dynamic graph drawing. IEEE Transactions on Visualization and Computer Graphics 14, 727–740 (2008)
43. Frishman, Y., Tal, A.: Dynamic drawing of clustered graphs. In: Proc. of the IEEE Symposium on Information Visualization, pp. 191–198. IEEE Computer Society, Washington, DC (2004)
44. Fruchterman, T.M.J., Reingold, E.M.: Graph drawing by force-directed placement. Software - Practice and Experience 21(11), 1129–1164 (1991)
45. Furnas, G.W.: Generalized fisheye views. In: Human Factors in Computing Systems CHI, pp. 16–23 (1986)

46. Gajer, P., Kobourov, S.G.: GRIP: Graph drawing with intelligent placement. In: Marks, J. (ed.) GD 2000. LNCS, vol. 1984, pp. 222–228. Springer, Heidelberg (2001)
47. Gamma, E., Helm, R., Johnson, R., Vlissides, J.: Design Patterns: Elements of Reusable Object-Oriented Software. Addison-Wesley (1994)
48. Gansner, E.R., Hu, Y., North, S.C.: A maxent-stress model for graph layout. In: Proc. of the IEEE Pacific Visualization Symposium, pp. 73–80 (2012)
49. Görg, C., Birke, P., Pohl, M., Diehl, S.: Dynamic graph drawing of sequences of orthogonal and hierarchical graphs. In: Pach, J. (ed.) GD 2004. LNCS, vol. 3383, pp. 228–238. Springer, Heidelberg (2005)
50. Görke, R., Hartmann, T., Wagner, D.: Dynamic graph clustering using minimum-cut trees. In: Dehne, F., Gavrilova, M., Sack, J.-R., Tóth, C.D. (eds.) WADS 2009. LNCS, vol. 5664, pp. 339–350. Springer, Heidelberg (2009)
51. Görke, R., Maillard, P., Staudt, C., Wagner, D.: Modularity-driven clustering of dynamic graphs. In: Festa, P. (ed.) SEA 2010. LNCS, vol. 6049, pp. 436–448. Springer, Heidelberg (2010)
52. Hachul, S.: A Potential-Field-Based Multilevel Algorithm for Drawing Large Graphs. Ph.D. thesis, Universität zu Köln (2002)
53. van Ham, F., Perer, A.: Search, Show Context, Expand on Demand: Supporting Large Graph Exploration with Degree-of-Interest. IEEE Transactions on Visualization and Computer Graphics 15(6), 953–960 (2009)
54. Harel, D., Koren, Y.: Graph drawing by high-dimensional embedding. In: Goodrich, M.T., Kobourov, S.G. (eds.) GD 2002. LNCS, vol. 2528, pp. 207–219. Springer, Heidelberg (2002)
55. Harrower, M.: Tips for designing effective animated maps. Cartographic Perspectives 44, 63–65 (2003)
56. Heer, J., Boyd, D.: Vizster: visualizing online social networks. In: Proc. of the IEEE Symposium on Information Visualization, pp. 32–39 (2005)
57. Hoogendorp, H., Ersoy, O., Reniers, D., Telea, A.: Extraction and visualization of call dependencies for large C/C++ code bases: A comparative study. In: Proc. ACM VISSOFT, pp. 137–145 (2009)
58. Hu, Y., Gansner, E.R., Kobourov, S.G.: Visualizing graphs and clusters as maps. IEEE Computer Graphics and Applications 30(6), 54–66 (2010)
59. Hu, Y., Kobourov, S.G., Veeramoni, S.: Embedding, clustering and coloring for dynamic maps. In: Proc. of the IEEE Pacific Visualization Symposium, pp. 33–40 (2012)
60. Inselberg, A.: Parallel Coordinates: Visual Multidimensional Geometry and Its Applications. Springer (2009)
61. Jaramillo, C.M.Z., Gelbukh, A., Isaza, F.A.: Pre-conceptual schema: A conceptual-graph-like knowledge representation for requirements elicitation. In: Gelbukh, A., Reyes-Garcia, C.A. (eds.) MICAI 2006. LNCS (LNAI), vol. 4293, pp. 27–37. Springer, Heidelberg (2006)
62. Joia, P., Paulovich, F.V., Coimbra, D., Cuminato, J.A., Nonato, L.G.: Local affine multidimensional projection. IEEE Transactions on Visualization and Computer Graphics 17, 2563–2571 (2011)
63. Kamada, T., Kawai, S.: An algorithm for drawing general undirected graphs. Inf. Process. Lett. 31(1), 7–15 (1989)
64. Kim, N.W., Card, S.K., Heer, J.: Tracing genealogical data with timenets. In: Proc. of the International Conference on Advanced Visual Interfaces, AVI 2010, pp. 241–248. ACM, New York (2010)

65. Koren, Y., Carmel, L., Harel, D.: ACE: A fast multiscale eigenvectors computation for drawing huge graphs. In: Proc. of the IEEE Symposium on Information Visualization, pp. 137–145 (2002)
66. Kumar, G., Garland, M.: Visual exploration of complex time-varying graphs. IEEE Transactions on Visualization and Computer Graphics 12(5), 805–812 (2006)
67. Lanza, M., Marinescu, R.: Object-Oriented Metrics in Practice - Using Software Metrics to Characterize, Evaluate, and Improve the Design of Object-Oriented Systems. Springer (2006)
68. Lyons, K.A.: Cluster busting in anchored graph drawing. In: CASCON, pp. 7–17 (1992)
69. Mashima, D., Kobourov, S.G., Hu, Y.: Visualizing dynamic data with maps. IEEE Transactions on Visualization and Computer Graphics 18(9), 1424–1437 (2012)
70. Mens, T., Demeyer, S.: Software Evolution. Springer (2008)
71. Moody, J., McFarland, D., Bender-DeMoll, S.: Dynamic network visualization. American Journal of Sociology 110(4), 1206–1241 (2005)
72. Moreta, S., Telea, A.: Multiscale visualization of dynamic software logs. In: Proc. Eurovis, pp. 11–18 (2007)
73. Moscovich, T., Chevalier, F., Henry, N., Pietriga, E., Fekete, J.D.: Topology-Aware Navigation in Large Networks. In: SIGCHI Conference on Human Factors in Computing Systems, pp. 2319–2328 (2009)
74. Muelder, C., Ma, K.L.: Rapid graph layout using space filling curves. IEEE Transactions on Visualization and Computer Graphics 14(6), 1301–1308 (2008)
75. Muelder, C., Ma, K.L.: A treemap based method for rapid layout of large graphs. In: Proc. of the IEEE Pacific Visualization Symposium, pp. 231–238 (2008)
76. Muelder, C.W., Crnovrsanin, T., Ma, K.L.: Egocentric storylines for visual analysis of large dynamic graphs. In: Proc. of 1st IEEE Workshop on Big Data Visualization (BigDataVis), pp. 56–62 (October 2013)
77. Xkcd #657: Movie narrative charts (December 2009), http://xkcd.com/657
78. Noack, A.: An energy model for visual graph clustering. In: Liotta, G. (ed.) GD 2003. LNCS, vol. 2912, pp. 425–436. Springer, Heidelberg (2004)
79. North, S.C.: Incremental layout in DynaDAG. In: Brandenburg, F.J. (ed.) GD 1995. LNCS, vol. 1027, pp. 409–418. Springer, Heidelberg (1996)
80. Ogawa, M., Ma, K.L.: Software evolution storylines. In: Proc. of the International Symposium on Software Visualization (SoftVis 2010), pp. 35–42. ACM, New York (2010)
81. Ogievetsky, V.: Plotweaver xkcd/657 creation tool (March 2009), https://graphics.stanford.edu/wikis/cs448b-09-fall/FPOgievetskyVadim
82. Orso, A., Jones, J., Harrold, M.J.: Visualization of program-execution data for deployed software. In: Proc. ACM SOFTVIS, pp. 67–75 (2003)
83. Paulovich, F., Eler, D., Poco, J., Botha, C., Minghim, R., Nonato, L.G.: Piece wise Laplacian-based projection for interactive data exploration and organization. Computer Graphics Forum 30(3), 1091–1100 (2011)
84. Paulovich, F.V., Nonato, L.G., Minghim, R., Levkowitz, H.: Least square projection: A fast high-precision multidimensional projection technique and its application to document mapping. IEEE Transactions on Visualization and Computer Graphics 14(3), 564–575 (2008)

85. Paulovich, F.V., Silva, C., Nonato, L.G.: Two-phase mapping for projecting massive data sets. IEEE Transactions on Visualization and Computer Graphics 16, 1281–1290 (2010)
86. Pfleeger, S.L., Atlee, J.M.: Software Engineering: Theory and Practice, 4th edn. Prentice Hall (2009)
87. Pretorius, A., van Wijk, J.: Visual inspection of multivariate graphs. Computer Graphics Forum 27(3), 967–974 (2008)
88. Purchase, H., Samra, A.: Extremes are better: Investigating mental map preservation in dynamic graphs. In: Stapleton, G., Howse, J., Lee, J. (eds.) Diagrams 2008. LNCS (LNAI), vol. 5223, pp. 60–73. Springer, Heidelberg (2008)
89. Reda, K., Tantipathananandh, C., Johnson, A., Leigh, J., Berger-Wolf, T.: Visualizing the evolution of community structures in dynamic social networks. Computer Graphics Forum 30(3), 1061–1070 (2011)
90. Robertson, G., Fernandez, R., Fisher, D., Lee, B., Stasko, J.: Effectiveness of animation in trend visualization. IEEE Transactions on Visualization and Computer Graphics 14, 1325–1332 (2008)
91. Rufiange, S., McGuffin, M.J.: DiffAni: Visualizing dynamic graphs with a hybrid of difference maps and animation. IEEE Transactions on Visualization and Computer Graphics 19(12), 2556–2565 (2013)
92. Rumbaugh, J., Jacobson, I., Booch, G.: The Unified Modeling Language Reference Manual, 2nd edn. Addison-Wesley (2004)
93. Saffrey, P., Purchase, H.: The "mental map" versus "static aesthetic" compromise in dynamic graphs: A user study. In: Proc. of the 9th Australasian User Interface Conference (AUIC2008), pp. 85–93 (2008)
94. Saha, B., Mitra, P.: Dynamic algorithm for graph clustering using minimum cut tree. In: SDM, pp. 581–586. SIAM (2007)
95. Sallaberry, A., Muelder, C., Ma, K.-L.: Clustering, visualizing, and navigating for large dynamic graphs. In: Didimo, W., Patrignani, M. (eds.) GD 2012. LNCS, vol. 7704, pp. 487–498. Springer, Heidelberg (2013)
96. Schaeffer, S.E.: Graph clustering. Computer Science Review 1(1), 27–64 (2007)
97. Simonetto, P., Auber, D., Archambault, D.: Fully automatic visualisation of overlapping sets. Computer Graphics Forum 28(3), 967–974 (2009)
98. Skupin, A., Fabrikant, S.I.: Spatialization methods: a cartographic research agenda for non-geographic information visualization. Cartography and Geographic Information Science 30, 95–119 (2003)
99. Tanahashi, Y., Ma, K.L.: Design considerations for optimizing storyline visualizations. IEEE Transactions on Visualization and Computer Graphics 18(12), 2679–2688 (2012)
100. Telea, A., Voinea, L.: An interactive reverse engineering environment for large-scale C++ code. In: Proc. ACM SOFTVIS, pp. 67–76 (2008)
101. Telea, A., Voinea, L., Sassenburg, H.: Visual tools for software architecture understanding: A stakeholder perspective. IEEE Software 27(6), 46–53 (2010)
102. Tufte, E.R.: Envisionning Information. Graphics Press (1990)
103. Wise, J.A., Thomas, J.J., Pennock, K., Lantrip, D., Pottier, M., Schur, A., Crow, V.: Visualizing the non-visual: spatial analysis and interaction with information from text documents. In: Proc. of the IEEE Symposium on Information Visualization, pp. 51–58 (1995)

9

Heterogeneous Networks on Multiple Levels

Falk Schreiber, Andreas Kerren, Katy Börner, Hans Hagen, and Dirk Zeckzer

At any moment in time, we are driven by and are an integral part of many interconnected, dynamically changing networks. Our species has evolved as part of diverse ecological, biological, social, and other networks over thousands of years. As part of a complex food web, we learned how to find prey and to avoid predators. We have created advanced socio-technical environments in the shape of cities, water and power networks, streets, and airline systems. In 1969, people started to interlink computers leading to the largest and most widely used networked communication infrastructure in existence today: the Internet.

Often, the complex structure of networks is influenced by system-dependent local constraints on the node interconnectivity. Node characteristics may vary over time and there might be many different types of nodes. The links[1] between nodes might be directed or undirected, and might have weights and/or additional properties that might change over time. Many natural systems never reach a steady state and non-equilibrium models need to be applied to characterize their behavior.

As a result of this usually domain-specific complexity, analysts are not only confronted with large multivariate networks. In practice, those networks can be assigned to different *levels* (or *scales*), and it is absolutely possible that several different networks share the same level. In case a set of networks share the same level, several notions of those networks can be found in the literature: they reach from *multimodal networks*, *network of networks* to *heterogeneous networks*. For simplicity, we use the latter term for the remainder of this chapter.

Complex structures of heterogeneous networks distributed over various levels lead to considerable visualization and analysis problems. First, the possibly tremendous size and complexity (in terms of higher dimensionality of the node/edge attributes) of those networks form a challenge of itself that is also

---

[1] We use the terms *link* and *edge* interchangeably.

discussed by other chapters of this book. Second and more important for this chapter, there are versatile relationships between the networks and/or between network elements across the different levels. In many application fields, it is essential to get a detailed understanding of such structures. In systems biology, for instance, networks are *the* key concept to structure and combine data that was generated by so-called high-throughput analysis methods. They can be arranged within a hierarchy of levels, from molecular-biological networks to evolutionary networks. The molecular-biological level, for example, contains a set of heterogeneous networks such as metabolic and gene regulatory networks. For a biologist, it is interesting to see how the different elements in those networks are connected with each other. Here, also multivariate data plays an important role as the network elements carry additional multidimensional information, such as experimental data that changes over time.

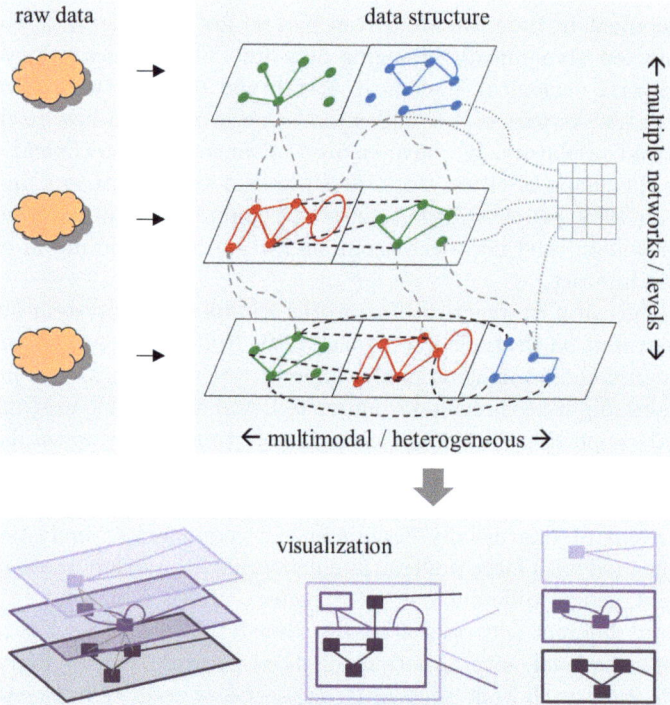

Fig. 9.1. An overview of the various topics of this chapter. Colors in the top right part are used to separate different graphs (e. g., within one level); colors in the bottom part separate different levels. Note that some nodes in the upper part have attached additional attributes symbolized by the small grey data matrix on the right hand side.

In terms of visualization, we want to navigate and explore through this world of networks, and a visualization tool should also provide techniques for the visual analysis of the multivariate data sets together with the underlying network topology. The upper part of Fig. 9.1 provides an overview of these concepts.

Many interactive visualization approaches and tools have been developed for the visual analysis of graphs/networks; the same is true for the visualization of multivariate data sets. We do not give a comprehensive overview of the literature here. Instead, we refer to textbooks, such as [38, 69, 70], surveys [24, 37, 46] and the introduction chapter of this book (Chapter 1). Note that the given literature references only point to selected example works and make no claim to be complete. However, a number of visualizations were especially developed for the analysis of multivariate networks [31]. Some of them are based on coordinated views, such as Jigsaw [67] or enRoute [56], others provide integrated solutions, for instance MobiVis [62], GEOMI [17], Vanted [60], or ViNCent [35, 72], and there are also approaches which realize so-called attribute-driven topologies like JauntyNets [32] and GraphDice [6]. We refer the reader back to Chapter 1 for a more detailed discussion of existing directions in multivariate network visualization. To the best of our knowledge there are no efficient visualization approaches for heterogeneous networks distributed over various levels. The aim of this chapter is to formalize and highlight the underlying problems and challenges by means of three application domains as well as to propose different solutions.

The remainder of this chapter is organized as follows: Sect. 9.1 provides a more formal specification of the data structures used in the chapter. Then, Sect. 9.2 exemplifies the visualization and analysis challenges by means of three important application fields: biology, social sciences, and software engineering. As there are almost no efficient visualization tools for multiple networks over levels available, we provide some ideas and visualization challenges in Sect. 9.3 (cf. the lower part of Fig. 9.1).

9.1 Formal Description of Used Data Structures

In order to facilitate the understanding and description of the following sections, we introduce a formal specification of the data structures used (a complete data structure sample is shown in Fig. 9.2). For this, let $G = (V, E, L)$ be a labeled graph with a finite set of nodes[2] $V = \{v_1, \ldots, v_n\}$, a finite set of edges $E = \{(v_i, v_j) | v_i, v_j \in V\}$, and a finite set of node and edge labels $L = \{l_1, \ldots, l_p\}$. Each node or edge of the graph is required to have a not necessarily unique label, and $l : V, E \to L$ gives the label for each node or edge. The label is used to encode a vector of additional node-/edge-specific data (i.e., the multivariate attributes).

---

[2] Often called *vertices*; therefore, the variable name v has been established to describe a node.

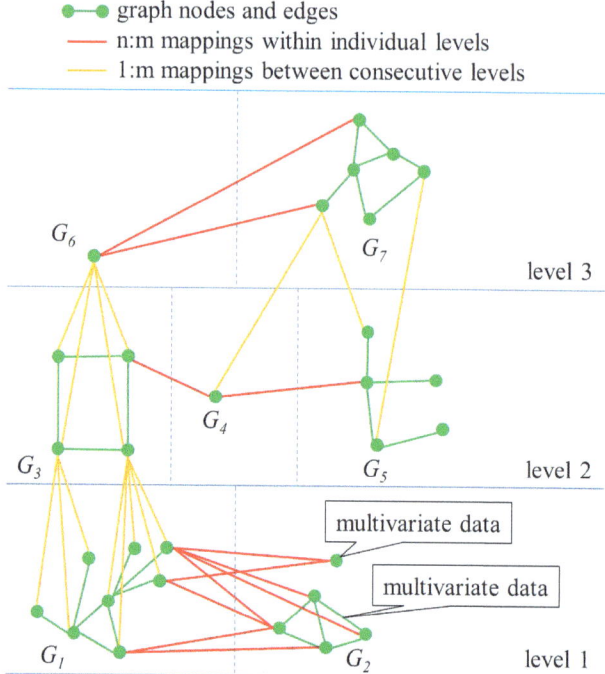

Fig. 9.2. The data structure described in Sect. 9.1. It shows three levels with two heterogeneous networks G_1, G_2 in level 1, three heterogeneous networks G_3, G_4, G_5 in level 2, and two heterogeneous networks G_6, G_7 in level 3.

Let $G_1, \ldots, G_m$ be a set of labeled graphs with $G_i = (V_i, E_i, L_i)$. Each graph may be directed, undirected, or mixed, and represents a specific network of the application domain (note that a graph might also be unconnected). For example, in the biological domain (see Sect. 9.2.1) a gene regulatory network may be represented by a directed graph, a protein interaction network by an undirected graph, and a metabolic network by a directed bipartite graph, respectively (cf. the green colored graphs in Fig. 9.2).

As mentioned in the introduction, we allow networks arranged at various levels. For modeling this property, let $S = \{s_1, \ldots, s_k\}$ be a set of consecutive levels. Each level can contain several graphs of $G_1, \ldots, G_m$, but each graph belongs to only one level. A level therefore groups graphs. The function $s : G, V, E \to S$ gives the level for each graph, node, or edge.

To connect graphs with each other, we introduce mappings between nodes of different graphs. Let $M = \{(v_i, v_j) | v_i \in V_i, v_j \in V_j\}$ be a mapping which connects a node from graph G_i with one node from a different graph G_j. Note that mappings within a graph are not allowed (those "intragraph mappings" are represented by the normal edges). Furthermore, the resulting structure could be seen as a new (global or union) graph G_G which merges all nodes and edges from $G_1, \ldots, G_m$ and the mappings (edges) M.

For simplicity, the mapping M will be restricted in the following way. For $m = (v_i, v_j)$ with $v_i \in V_i$ and $v_j \in V_j$: if both nodes belong to the same level—$s(v_i) = s(v_j)$; see the red links in the figure—there will be no restriction. However, if both nodes belong to different levels (yellow edges), a mapping is only allowed if the following two conditions hold:

1. both levels are consecutive (neighboring) levels, i.e., $s(v_i) = s(v_j) - 1$ or $s(v_j) = s(v_i) - 1$, and
2. there is a $1 : n$ mapping from the higher to the lower level—i.e., if $s(v_i)$ is the higher level ($s(v_i) = s(v_j) + 1$) then there is no other node v_k in level $s(v_i)$ with $(v_k, v_j) \in M$.

9.2 Application Domains

This section provides an overview of the most important visualization and analysis challenges by means of three application fields: life sciences / biology, social sciences, and software engineering.

9.2.1 Life Sciences / Biology

Biological processes are commonly represented as networks. Examples of biological networks are molecular biological networks such as protein interaction networks (showing the interaction possibilities of proteins) and metabolic pathways (representing the transformation of metabolites into other metabolites), food webs and ecological networks (showing the dependencies between prey and predators), and phylogenetic networks (representing the evolutionary relationships between species).

In the remaining of this section we first present a domain overview introducing relevant terminology along the way. The next subsection discusses major data sources and formats. Then we illuminate diverse network types and interlinkages together with examples. The next subsection presents concrete use cases that benefit from the formalization presented in the previous section and the visualization solutions discussed later. We conclude with a discussion of challenges for multi-level network analysis and visualization in the life / biological sciences.

Overview

A better understanding of biological networks helps in making sense of much of the complex data which is nowadays available in biology, biochemistry, medicine, and related areas of the life sciences. Visualization is a key method to foster exploration and understanding, and, therefore, biological network visualizations have existed for a long time. The importance of visualization and visual analysis of these networks is also evidenced by the large number of books, tools and databases that either contain manually produced drawings of biological processes and networks, or provide algorithms for their automatic layout. Many tools are available, some comparisons of tools have been presented (for example, in [22, 43, 60]), and a number of well known tools supporting network visualization and analysis are:[3]

- *BiNa* [45]   (http://bit.ly/y6ix9i)
- *BioUML* [42]   (http://bit.ly/yIETIt)
- *CellDesigner* [20]   (http://bit.ly/AOFQiF)
- *CellMicrocosmos* [66]   (http://bit.ly/WJ8cnE)
- *Cytoscape* [65]   (http://bit.ly/wY2sbG)
- *Ondex* [41, Chapt. 5]   (http://bit.ly/AetZjz)
- *Pathway Projector* [43]   (http://bit.ly/zo5x2M)
- *PathVisio* [27]   (http://bit.ly/zunwxW)
- *SBGN-ED* [13]   (http://bit.ly/17m7KfW)
- *Vanted* [30]   (http://bit.ly/Aigr0T)
- *VisAnt* [25]   (http://bit.ly/agZBni)

Data Sources

Biological networks may be directly derived from experimental data (such as protein interaction networks) or are built based on knowledge (such as metabolic networks). There are many data sources for biological networks: from databases covering a specific domain for a specific species (such as Ara-Cyc [54] for metabolism in Arabidopsis) to a specific domain for a set of species (such as MetaCrop [61] for metabolism in crop plants) to several domains for several species (such as KEGG [34] for metabolic and signalling processes in a wide range of species). Another important criterion is the quality of the data which can range from completely manually curated, high-quality data to computationally derived, uncurated data. For overviews of databases for a range of data domains, see [4, 18], for instance.

To support exchange between tools and databases, a few standard representations are widely used such as SBML [26] and BioPAX [14]. Also the graphical representation of cellular processes and biological networks has been standardized with SBGN [48] which helps in understanding the complex processes due to the unambiguous use of glyphs. For details, see the specifications of the SBGN languages [51, 53, 55].

---

[3] Note that there are more than 170 tools available for network visualization and analysis, and a complete listing and comparison is beyond the scope of this article. We list some tools here which exists since several years and often allow easy extensions via plugin mechanisms.

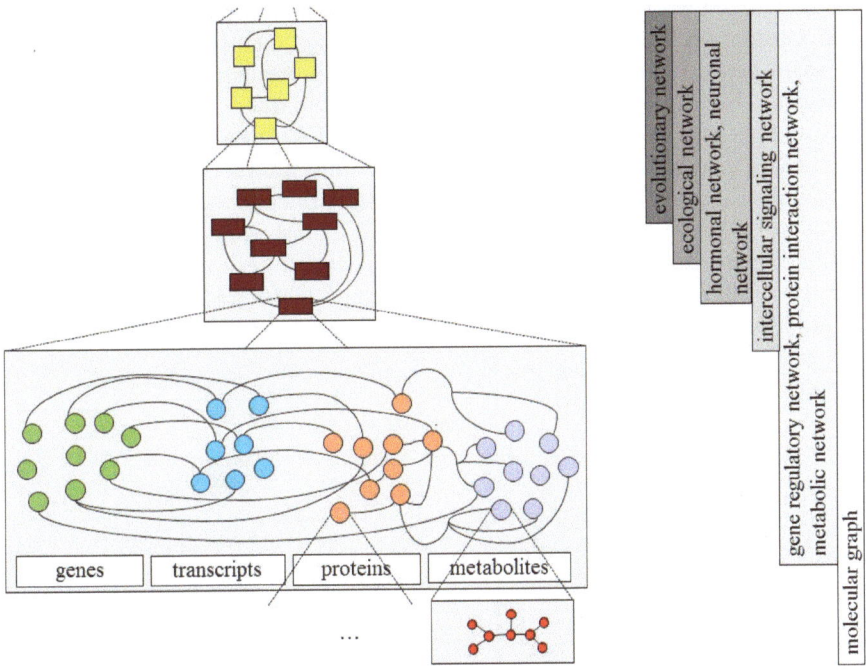

Fig. 9.3. A hierarchy of biological networks

Network Types and Examples

Several types of graphs are used to represent biological networks and some typical examples are presented in the following. *Directed graphs*: gene regulatory and signaling networks which describe how genes can be activated or repressed and therefore which proteins are produced in a cell at a particular time; food webs which model the relationships between species in an ecological system. *Undirected or mixed graphs*: protein interaction networks which represent the interaction between proteins such as the building of protein complexes and the activation of one protein by another protein. *Hypergraphs or bipartite graphs*: metabolic networks which show how metabolites are transformed—for example, to produce energy or synthesize substances. *Trees*: phylogenetic trees which are commonly built on information from molecular biology such as DNA or protein sequences and which represent the ancestral relationships between different species. There exists a *hierarchy of biological networks* (see Fig. 9.3 and also Chap. 4).

An example of the data structure described in Sect. 9.1 is given in Fig. 9.4. Here, three molecular-biological networks (gene-regulatory network, protein interaction network, and metabolic network) are presented on level 1. Genes in the gene regulatory network may activate or inactivate the transcription of other genes. Genes are transcribed into proteins, therefore the protein interaction network does not only contain edges between nodes of the protein interaction network (for representing interaction), but also (red) edges between genes (gene regulatory network) and proteins (protein interaction network). Finally the metabolic network is a bipartite graph consisting of metabolites (circles) and enzymes (rectangles). Enzymes are proteins; therefore, there are (red) edges connecting the protein interaction graph with the metabolism graph. On the next level, two networks are represented: an association of all genes to their chromosomes and a clustering of all proteins into disjunct clusters. The yellow edges represent which gene of the gene regulatory network belongs to which chromosome, and which protein of the protein interaction network belongs to which protein cluster, respectively.

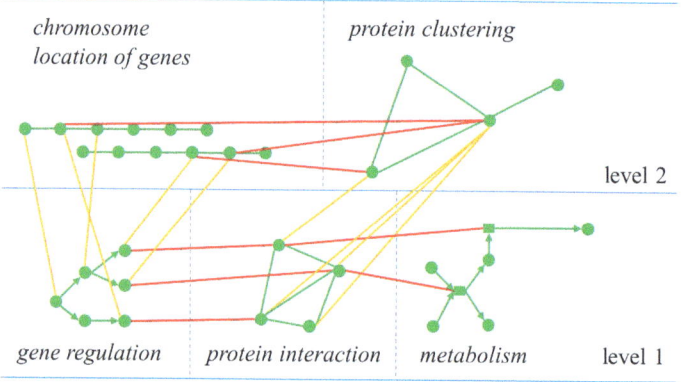

Fig. 9.4. An example instantiation of the data structure described in Sect. 9.1

Typically multivariate data is connected to nodes and/or edges of the networks. Figure 9.5 gives some examples of such data. Here we will focus on the network structure, but it should be kept in mind that not only the visualization of the networks, but also of the additional multivariate data is often a challenge.

Use Cases

Here we discuss use cases that involve the multi-level heterogeneous network shown in Fig. 9.4. Genes encode proteins, and proteins have many functions including catalyzing metabolic reactions in the form of enzymes. This major flow of information motivates the use of the three networks on level 1 in Fig. 9.4: gene regulatory, protein interaction, and metabolic network. These networks focus on different aspects of the same underlying biological system. In addition, level 2 gives further information which is either experimentally obtained (such as the location of genes on the chromosomes) or computed

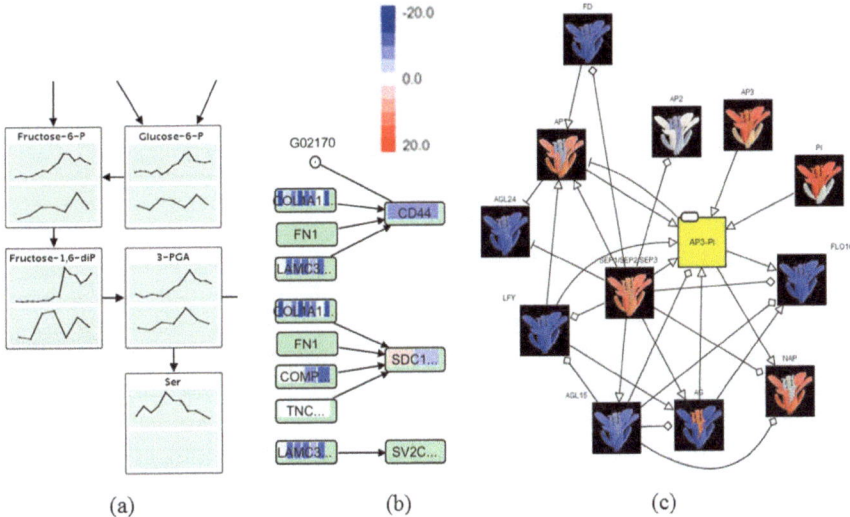

Fig. 9.5. Examples of multivariate data in biological networks, (a) time-series of relative metabolite levels in a metabolic network (for two different conditions: day and night; from [7]); (b) up- (red) and down-regulated (blue) genes in a signalling pathway (from [39]), (c) spatial resolution of gene expression in the gene regulatory networks of Arabidopsis (from [29]).

(such as the clustering of proteins based on connectivity information).[4] Typical tasks involving these networks are:

Structural properties of the networks

Which elements of the networks are important nodes (such as regulatory genes), functional building blocks (such as feed-forward motifs [52]), or relevant paths through the network (such as the shortest path between two metabolites)? Such questions are usually answered using methods from network analysis: centrality analysis to evaluate the importance of nodes or edges in the network [44], network clustering which may structure the network into functional modules [5], shortest paths between nodes representing potentially preferred routes, and so on. The results of such analysis have to be visualized in the network context and may give new insights into how important a specific gene is for an organism or which metabolic pathway may be preferred by an organism.

Networks and spatial information

Are genes with close proximity on the chromosome involved in the same biological processes? For example, if genes in close proximity on the chromosome

---

[4] Note spatial location and aggregation or grouping of elements can also occur in other application domains; for example, see Fig. 9.6.

are regulated by the same regulators in the gene regulatory network or if they fall in the same cluster in the protein clustering, then they are likely to belong to the same biological process.

Networks and multivariate data

Is experimental data in agreement with knowledge represented by the networks or not? Lots of experimental data can be attached to nodes and edges: data obtained under different experimental conditions such as treatments or temperature, time-series measurements, and so on. For example, the following attributes can be used: edge weights (e. g., to represent fluxes), node weights (e. g., to show concentrations), node presence/absence (e. g., to model knockouts), and node shape (e. g., to show different types of biological objects). Multivariate data has first to be integrated into the biological network. Data mapping deals with this integration of additional data into networks. An example is the integration of metabolomics, transcriptomics, and fluxomics measurement data into the metabolic network. This data can be mapped on different network elements (such as metabolites, enzymes, and reaction edges). The mapped data should behave in a way which could be explained by the underlying networks. If, for example, in the gene regulatory network, only some of the dependent genes get activated if a regulatory gene is active, this could be an indication for a yet undiscovered regulatory process.

Challenges

The compilation of heterogeneous networks requires the identification of the biological entities such as genes, proteins, metabolites, species and so on, and the interconnection between the networks with (different types of) edges. The biological entities are only partly known; therefore, the networks are not complete but change with increasing knowledge. Also, the interconnection between the heterogeneous networks is often difficult to obtain: identifiers for biological entities are often only unique in the context of one data source—for exmple, a database or an ontology. There are initiatives aiming towards the creation of globally unique and persistent URIs (e. g., the MIRIAM registry [47] and identifiers.org [33]), but they are still only used in some databases. A commonly used approach is identifier mapping (see [50] for an overview of tools capable of mapping biological database identifiers).

Tools for the visualization of analysis results often provide standard interaction techniques such as zooming (sometimes also semantic zooming), filtering, collapsing and expanding of structures, and highlighting. There are many tools available, but choosing the right tool is often difficult (for links to some more widely used tools, see the introduction of this section). Standard visualization tasks are often sufficiently solved, but more elaborated visualization, interaction, analytics, and layout methods are an open issue.

An overview concerning open problems in biological network visualization can be found in [3].

9.2.2 Social Science

This section reviews heterogeneous, multi-level networks in the social sciences, drawing on research in sociology, information science, statistics, graph theory, and network science [9]. A special focus will be scholarly networks such as citation networks of papers or journals, and collaboration networks of authors, their institutions, and countries. We first present a domain overview introducing relevant terminology along the way. The next subsection discusses major data sources and formats. Then we illuminate diverse network types and interlinkages together with examples. In the next subsection we present concrete use cases that benefit from the formalization presented in Sect. 9.1 and the visualization solutions discussed in Sect. 9.3. We conclude with a discussion of challenges for multi-level network analysis and visualization in the social sciences.

Overview

Social networks exist from the individual (micro) to population (macro) levels. Examples are networks of friendship and hate, collaboration and competition, and trade and blockage between entity nodes such as individuals, organizations, cities, countries, geospatial regions, or areas of science. Typically, an entity node is part of multiple types of networks—for example, a person is part of friendship, collaboration, and family relationship networks. Nodes might contribute to information diffusion networks by receiving and sending emails, tweets, or commenting on digital objects. Nodes from different levels interact and influence each other—for example, individual authors of papers might be aggregated at the institution or even country level, see Fig. 9.6. Journal publications can be aggregated into journals. Undirected collaboration and directed citation edges are aggregated as well. Network edges can be mapped geospatially, supporting geospatial grouping by region, county, country, or continent (see lowest level in Fig. 9.6).

The movement of a highly cited author from one department or country to another will impact both the expertise profiles (entity properties) and collaboration patterns (entity linkages) of associated entity nodes. In other cases, networks from different levels are nested—for example, a person is part of local social networks that can be further aggregated to global, population level networks. Last but not least, the structure of multi-level networks impacts the utility of the network for, for example, information diffusion. The stronger the edge between two nodes is, the more information can flow; the more often two nodes share information, the stronger their edge grows over time.

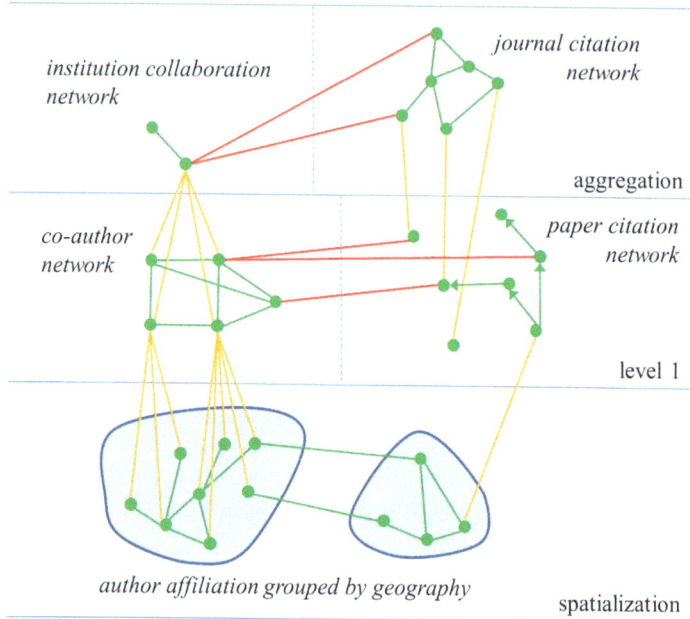

Fig. 9.6. Three-level network of different collaboration and citation networks

Understanding the evolution of network structure and dynamics has far-ranging applications. Among them are the prevention of disease spreading, for example, identifying highly connected individuals that should be vaccinated first when trying to fight a pandemic or using social networks to reducing the social diffusion of smoking or bad eating behaviors (see subsection "Use cases"); to increase the spreading of information—for example, in education or when designing effective (viral) marketing strategies; to help manage the extremely complex decision making space of professional career choices (see subsection "Use cases"); or to form and sustain productive research and development teams.

Data Sources

Social network data might be qualitative or quantitative. Qualitative data is commonly acquired via surveys, interviews, direct observation, or by reviewing written documents. Quantitative data might be derived from existing databases (e. g., phone-address-data-revealing social networks; publication data for extracting co-author networks) or acquired via measurement. Common data sources are social media data (phone, email, blogs, Twitter, Facebook), scholarly data (scientific publications, patents, funding awards), or custom data collected in research studies. Commonly studied networks are social networks, friendship networks, collaboration networks, political networks, trading networks, and citation and other knowledge diffusion networks.

Nodes and their linkages can be represented as an adjacency matrix, an edge list, or lists of nodes and edges. Consequently, most network science tools[5] support Pajek .net or edge list formats as well as Graph/ML, XGMML formats. Many tools support the extraction of networks from common file formats—for example, publication data formats from the Web of Science, Scopus or personal bibliography tools such as EndNote or Latex's bibtex, or from generic tabular formats (e. g., .csv files).

Network Types and Examples

This subsection discusses different network types using examples from scholarly network analysis—the study of authors and their papers as we assume all readers are familiar with this data.

Networks might be directed or undirected, weighted or unweighted, valued or not. Linkages might be among units of the same type, such as friendship or co-authorship linkages, or between units of different types, such as authors and the papers they produce. In general, three types of linkages are distinguished: *direct linkages* such as paper citation linkages; *co-occurrence linkages* of words or references; and *co-citation linkages* (e. g., of authors or papers). Plus, units of the same type can be interlinked via different link types: for instance, teenagers might be linked via love and hate relationships; papers can be linked according to co-word, co-citation, or bibliographic coupling analysis. Linkages might be directed and/or weighted. Each non-symmetrical occurrence matrix has two associated (symmetrical) co-occurrence matrices; for instance, for each paper citation matrix exists a bibliographic coupling and a co-citation matrix.

An example of direct linkages are paper-paper (citation) linkages: Papers cite other papers via references, forming a non-weighted, directed paper citation graph. It is beneficial to indicate the direction of information flow from older to younger papers via arrows. References enable readers to search the citation graph backward in time. Citations to a paper support the forward traversal of the graph. Citing and being cited can be seen as two vital roles of a paper.

Co-occurrence linkages interconnect co-author networks. Having the names of two authors (or their institutions and countries) jointly listed on one paper, patent, or grant is an empirical manifestation of scholarly collaboration. The more often two authors collaborate, the greater the weight of their joint coauthor link. Weighted, undirected co-authorship networks appear to have a high correlation with social networks that are themselves shaped by geographic proximity.

Word co-occurrence linkages are used to calculate the topic similarity of basic and aggregate units of science. Units that share more words are assumed

---

[5] http://sci2.wiki.cns.iu.edu/display/SCI2TUTORIAL/ 8.2+Network+Analysis+and+Other+Tools

to have higher topic overlap and are connected via linkages and/or placed in closer proximity on a topic map. Co-occurrence networks are weighted and undirected.

Co-citation linkages is as follows: Two basic or aggregate units of science are said to be co-cited if papers associated with them jointly appear in the list of references of a third paper. The more often two units are co-cited, the more they are expected to have something in common. Examples are document co-citation and author co-citation networks.

Given a data file with publication records retrieved from the Web of Science or Scopus database, more than 30 different networks can be extracted. Some of these networks—for example, co-author or paper-citation networks—have been studied extensively and their structure, distribution, and evolution are known. Other network types—particularly heterogeneous networks interlinking different levels—have not yet been studied in detail.

Use Cases

This subsection exemplarily discusses two use cases that involve multi-level heterogeneous social networks. For each, we identify user needs/tasks, workflows, and insights gained.

Reducing social contagion of smoking behavior

Social peer pressure is powerful. The desire of an individual to be an accepted member of a group frequently leads to behavior that the individual would not show without the group. For example, it is well known that— all other things being equal—joining a well-organized team of experts will lead to higher professional performance than joining a team of less skilled, less organized individuals. Studies by Christakis and Fowler have evaluated a densely interconnected social network of 12,067 people assessed repeatedly from 1971 to 2003 as part of the Framingham Heart Study to show the relevance of social networks in the diffusion of not only smoking and obesity [10], but also loneliness. Relevant news stories were entitled: "Are Your Friends Making You Fat?"[6] While it is difficult and in some cases impossible to distinguish correlation vs. causation [11]—for example, some of the shown effects could be due to the "birds of a feather flock together" factor also known as homophily—networks seem to impact outcomes. A scientific approach that considers homophily, environment, and induction (e. g., by using multi-level heterogeneous networks) seems promising.

For example, it appears beneficial to study individuals in the context of multiple, interconnected networks. Level 1 might comprise the different social networks of a teenager node A. Among them are family, school, and out-of-school friendship networks. Level 2 aggregates these networks to families and

---

[6] http://www.nytimes.com/2009/09/13/magazine/
13contagion-t.html?pagewanted=all&_r=0

clans (groups of teenagers) that have diverse profiles and interlinkages. Geography matters: the ease by which A can come into contact with teenagers and other individuals that have different demographic profiles is important—just like in Fig. 9.6 a spatialization level can be used to represent geospatial factors. Using this representation, the number of linkages to positive and negative influences can be calculated—for example, the strength of network ties to peers and family members, the rate of contact, and geospatial proximity can be determined for any individual node. Using computational models, it might be possible to predict the general impact of network changes on individual behavior. The big, open question is: How to change social networks and/or environments to cause positive change?

Designing successful career trajectories

Pathways that individuals choose from taking their very first job and then moving on to the next during their working years are called career trajectories. They might be plotted over time, geospatial space, or topic space. Individuals might change locations and jobs because of warfare, political problems, ethnic purging, or because of voluntary or volitional migration where individuals choose to relocate to new places because of opportunities offered in the new place. In the latter case, social factors (e. g., closeness to family and friends, standard of living), colored by cultural, historical, linguistic, or weather considerations, but also active encouragement by visa and immigration controls are key criteria. Education is another major factor—low education typically equates low paying jobs and little resources to pay for education or to move to a different place. High education commonly leads to better paying jobs and the generation of financial wealth that pays for personal education or the education of family members. Highly specialized expertise profiles might also mean that only a few jobs exist that match this expertise profile and international migration is required to find an appropriate job.

Coming back to the study of scholarly networks, much data exists to track career trajectories of scholars over time. The U.S. National Science Foundation has been conducting the Survey of Doctorate Recipients (SDR) biennially since 1973. The SDR follows a sample of U.S.-trained doctorates in science, engineering, and health fields throughout their careers from shortly after degree award by a U.S. institution through age 75. Multivariate data such as detailed information on professional position, salary, number of kids, etc. is available for each respondent in a longitudinal fashion. In addition, there exist extensive publication and funding databases that record funding intake (number of awards and funding amounts) and publication output (number of papers and their citations) over time together with information on not only co-investigators and coauthors, but also institutional affiliations, and acknowledged grant funding that interlinks funding and publications.

In general, career trajectories are best viewed as decisions over time. As it is much easier to change institutions than to change a topical area of expertise; researchers change geolocations frequently (particularly in the beginning of their career). However, they are less likely to change their topic area and, if they do, decide to venture into topically similar areas of science that benefit from the same skill set/expertise. Multi-level networks can be used to represent the impact of geospatial features—for example, car routes and air traffic networks as a proxy for reachability (spatialization level). Level 1 might show scholar node B and associated family links, mostly local friendship networks, (international) co-author relations, and collegial networks at the same institution. Level 2 might be used to represent the reputation and interlinkages of institutions and scientific disciplines. Family and friendship links, the reputation of another institution, and also attributes of the new geolocation (living costs, weather, etc.) have the power to make a person look for or accept another job—if a job offer is made. The new institutional environment together with new friendship and collaboration opportunities will change funding and publication patterns of an individual scholar and ultimately the conditions for the next career step. Note that not only individuals and families but also companies and organizations migrate in response to market changes, to maximize economic utility, or to be close to customers.

Challenges

The extraction of heterogeneous networks requires the identification of unique entities—for example, people, institutions, scientific areas, and their interlinkage via (different types of) edges. While publications, patents, and other digital documents have unique digital object identifiers, the development of digital object identifiers for authors, institutions, and scientific areas is still under active development. Similarly, the interlinkage of (heterogeneous) networks across levels poses serious unification and data mapping challenges. Many problems require $n:m$ mappings not only within one level but also between consecutive levels—implications for restricting the mapping to $1:m$ are unknown. For example, Fig. 9.6 assumes that each author has exactly one affiliation and each institution is mapped to one country—this is not true for all authors nor institutions.

Few algorithms exist to analyze and visualize heterogeneous networks—most algorithms assume there is one node type and one edge type. Visualizations are important to communicate complex heterogeneous, multi-level networks. However, most readers hardly ever learned how to read and interpret a network and seeing multiple interlinked networks with different node and edge types is often overwhelming.

9.2.3 Software Engineering

A plethora of tasks in software engineering involves analyzing software artifacts that are best represented using networks, see also Chap. 2. We will focus on object-oriented programming languages like Java or C++ for implementation. Other languages will lead to similar structures. The different networks can be categorized into static software structure (e. g., classes in Java or C++ and their connections) and dynamic software structures (e. g., which classes are instantiated and when and how these instances are connected to each other). Furthermore, different structural connections between the same nodes (classes) are possible—for example, call graph, inheritance (both directed) or code clones (undirected). This forms a within-level set of graphs. Another within-level set of graphs is given by the change of the graphs over time. During the evolution of the software system (seen as snapshots in repositories like svn, cvs, or git), the graph changes: nodes (classes) are added, deleted, or changed, and also links are added or deleted, or they change their properties (e. g., the number of methods of one class called by another one). Finally, software entities are structured: methods belong to classes belong to modules. These modules can be represented by, e. g., packages in Java or namespaces in C++. This package or namespace tree can be directly mapped to the levels of Fig. 9.2.

In the following of this section we first present a domain overview introducing relevant terminology along the way. The next subsection discusses major data sources and formats. Then we present diverse network types and inter-linkages together with examples. The following subsection presents concrete use cases. We conclude with a discussion of challenges for multi-level network analysis and visualization in software engineering.

Overview

The data produced and used in software engineering is manifold. First of all, software development processes can be mapped to directed graphs with loops, where each development step is mapped onto a node, and an edge designates the sequence of these steps. Further, software development artifacts like system designs, detailed designs, and the static and dynamic structure of programs can be mapped onto directed and undirected graphs. Finally, algorithms can be visualized. In this latter case, graphs are more rarely used, except that the algorithm requires the handling of graphs.

The networks can be time-dependent ($n : m$ mapping within a level)—for example, data from cvs, svn, or other repositories—and multi-level ($1 : n$ mapping between levels)—for example, package structure in Java. Further, they have different node and edge types, and to each of these nodes and edges additional information (multivariate data) can be associated. Thus, the networks have all properties described in Sect. 9.1 and shown in Fig. 9.2.

An exemplary instantiation of Fig. 9.2 is shown in Fig. 9.7. All nodes in the lowest-level—level 1—represent classes. These classes are related by edges (green) representing, for example, method calls. They are grouped according to their revision (release); three releases are shown. Classes (nodes) that are present in subsequent revisions are connected by red edges. Aggregating classes into packages (yellow edges) results in the graphs on level 2. Here, the

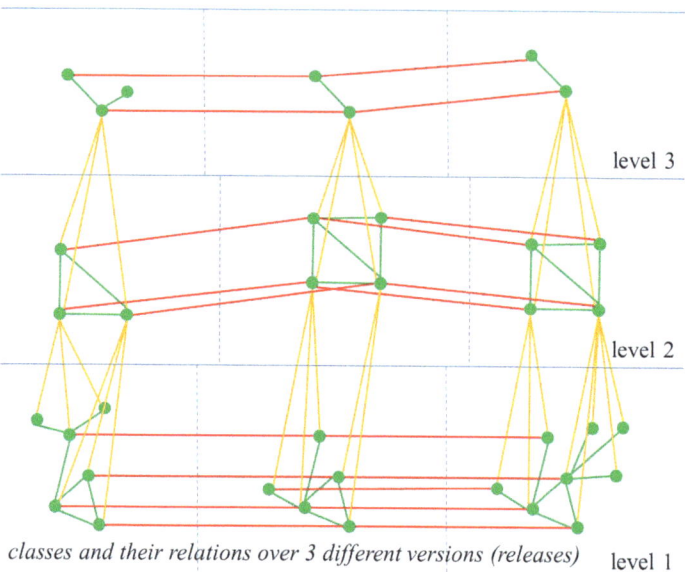

Fig. 9.7. An example instantiation of the data structure described in Section 9.1

nodes represent both nodes and packages. Packages are those nodes which are connected by yellow edges to level 1. The same holds for level 3.

Understanding software processes is important for assigning and managing resources during the software development process, while understanding the structure and the behavior of software products (software comprehension) is important for bug fixing and extending existing software systems (software maintenance).

Data Sources

Networks in software engineering are either created manually or automatically. Manual creation occurs normally for describing software engineering processes and during software development in the design phase. However, the latter networks are only reliable in the case of automatic code generation. Whenever code is generated manually, the design is usually changed either intentionally or unintentionally. Therefore, software systems are commonly analyzed based on graphs extracted directly from the source code. The source code itself comes from software repositories like svn, cvs, or git. During the extraction, the graphs are generated. Some systems additionally extract so-called software metrics. In the latter case, these software metrics are used as attributes of the classes (nodes). The attributes for the edges can also be inferred from these metrics. On the other hand, the number of edges of the same type can be counted and used as an additional measurement (e. g., how many different method calls occur from class A to class B).

The granularity of the analysis influences which are the lowest-level entities (nodes). Most often, either classes, methods or member variables of classes are chosen as atomic entities. Methods and member variables of classes are a part of classes, which in turn are a part of packages. Packages themselves form a tree with one root package or a forest with several root packages. Edges might represent directed relations like method calls, inheritance, instantiation, or aggregation, or undirected relations like code clones or common fate.

Finally, execution logs (traces) of program runs are mined to create dynamic call graphs. In this case, nodes are mostly the instantiated classes, while edges are mostly the dynamic call relations between the class instances. However, static relations like aggregation or inheritance can help to understand the behavior of the programs. More data sources and how these sources are mined to obtain information are described in Sect. 2.3.3 of this book.

Network Types and Examples

Software processes are mapped onto networks similarly to other processes (e. g., business processes, schedules, or production processes). Each development step is mapped onto a node and can have additional information assigned (e. g., number of developers, time allocated for this step, a list of tasks, inputs, and deliverables). Each step is connected to the following step by an edge. This edge normally is guarded in the sense that it can be only followed if the deliverables of the previous step are ready (to a certain amount). As the processes can branch and can have loops, the graphs cannot be reduced to trees or DAGs (directed acyclic graphs). Mostly, the process comes in an abstract form that is instantiated for a specific project. Therefore, several similar networks exist that can be compared to each other (e. g., the instantiations among each other). Further, the process can be in the form of a hierarchy. Then, each node of the high-level process is refined into a number of steps that can be refined themselves.

The static structure of software artifacts can also be mapped to typically directed graphs. Taking Java source code, these networks have the following properties. The hierarchy of Java packages can be mapped to different levels. The lowest-level nodes can be either classes or methods, members, and subclasses of classes. The snapshots in time (typically saved in revision-control systems) form a sequence of hierarchical networks. Many revision-control systems allow branches and merges, such that the sequence becomes a directed acyclic graph. Possible links (edges) on class level are inheritance, method call, aggregation, implements relation, and usage relation (similar to aggregation, where the class is used as type of another class' member, it can also be used as type of return values, parameters, or exceptions). Besides these directed links (edges), also undirected links (edges) might also exist like code clones, common fate (evolution, classes that are often changed together), or semantic similarity. Aggregating the classes and sub-packages in packages directly implies the aggregation of the edges; thus, a hierarchy is built for one time-step.

The dynamic structure of a software system also leads to a sequence of networks. However, these networks typically have a $1 : n$ mapping as the underlying static structure does not change. Instead, individual edges are activated and deactivated according to the dynamic evolution of the networks. Indeed, the nodes of the dynamic networks are instances of the nodes of the static network, where each static node can lead to an arbitrary number of dynamic nodes. The same holds for the edges: call graph edges get instantiated during the evolution of the run. Of course, the granularity can be again on class or on any of the package levels.

Use Cases

Use cases are typically directly derived from software engineering tasks (see also Sect. 2.3 in this book). Tasks that benefit from visualizations include

Software understanding

Understanding the functionality and the interplay of components necessitates the understanding of the static and dynamic structure. Further, it is the basis for the maintenance task.

Maintenance

To maintain a system, a thorough understanding of the individual components as well as their dependencies is needed. The dependencies are best analyzed based on visualizations of the static structure and dynamic runs.

Re-engineering

Re-engineering is needed for legacy systems or whenever the new developer can not directly access the knowledge of the original developers. Like Software understanding, the interplay of the different components plays an important role, which is best understood using static and dynamic networks.

Testing and bug fixing

For bug fixing, the interplay of components is very important, because side effects on other components should be minimized when changing an individual component. The same holds for adding new functionality. The current behavior should not change by adding additional functionality.

Product lines

Product lines incorporate basic functionality in a core of components, adding functionality by additional or changed components. For product lines, it is mandatory to understand the relationship between the different products. The goal is to maximize the amount of common components, minimizing the effort for creating additional functionality using additional components.

Challenges

There exist a lot of software visualizations, but only a few of them are scalable and comprehensive. Most of the existing solutions either focus on specific properties of software artifacts, like metrics, static structure (UML diagrams) at one time step with several types of edges, or static structure over time with one edge type.

Highly necessary for effective program comprehension are integrated views like the AreaView tool developed by Byelas at al. [8]. While this tool integrates UML diagrams with metrics and several areas of interests, it displays neither hierarchies nor several time steps. We need an integrated set of visualizations and interactions that allows us to look at static software artifacts from several points of view, showing or filtering information on demand to allow a focused analysis of the software artifacts. Such approaches should visualize networks with the following properties:

1. time-dependent static structures: e. g., svn or cvs snapshots;
2. structure hierarchies: e. g., packages and classes in Java;
3. distinct node types: e. g., classes and interfaces in Java;
4. distinct edge types: e. g., inheritance, aggregation, method calls (directed) and code clones, semantic similarity (undirected);
5. software metrics as additional node information: e. g., lines of code, number of methods, depth of inheritance;
6. metrics associated with edges: e. g., number of method call relations, similarity.

While all of these properties mainly map to visual structures, interaction has always to be considered as part of the solution—that is, at least the use of standard interaction techniques is mandatory. Only if users are allowed to interactively explore selected parts of the software artifacts, they will be able to gain new insights and find the information needed for solving specific tasks. Thus, interactive visualizations must be seamlessly embedded in the software analyst's work flow. Only then can analysts discover complex patterns in software. Specific challenges derived from this general challenge are:

Scalability: How to depict several levels of several revisions over time? Usually, the graphs are so large that a cut through the level hierarchy is needed showing the focused information in detail (lowest-level) and the context information reduced (on a higher level). Showing a series of cuts is then asked for to analyze the evolution of the software system for finding patterns or unusual changes, see also Chap. 10.

Comprehension: What is the best visualization-interaction combination for showing all relevant information?

- *Metrics:* On the one hand, metrics (i. e., attributes on the nodes and edges) have to be included in a non-obstructive way with respect to the structure.
- *Multiple Edge Types:* On the other hand—besides the hierarchy—each level might also represent different edge types. Then the yellow edges in Fig. 9.7 represent again the same node, while the levels in this case represent different structural information like call graphs, inheritance, or code clones. Possible solutions for this problem have already been proposed by Abuthawabeh et al. [1, 2] based on a matrix visualization. A similar approach using node-link diagrams has been proposed by Knodel et al. [40]. However, these two approaches need further improvements to become valuable in software comprehension tasks.

9.3 Visualization

The visualization of heterogeneous networks on multiple levels is still relatively unexplored in the literature. However, there are a number of visualization approaches that focus on solving specific analysis tasks or operate on a subset of the data structure introduced in Sect. 9.1, for instance on heterogeneous multivariate networks: some approaches abstain from explicit encoding of the network topology and visualize aggregated information only. This idea supports the analysis of very large data sets in terms of many heterogeneous networks and large multivariate attributes. ManyNets [19] represents networks as rows in a table together with their multivariate data (primary as well as secondary data [37]) similar to the well-known TableLens [58]. Several interaction possibilities support the visual analysis of the networks which might also be displayed as node-link diagram on demand. GraphTrail [16] has similar aims, but in contrast to showing the networks in a table, the developers have chosen to represent the network elements in an aggregated form. For doing this, standard charts like bar charts or tag clouds are employed that can be interactively arranged on a canvas. GraphTrail also supports the analysis process by providing a history functionality (Sect. 6.3.2 in Chap. 6 discusses GraphTrail in more detail). Other approaches abstract directly in the node-link representations, such as OntoVis [63]: an ontology graph, which describes the node categories/clusters and their relationships and serves as a vehicle to control the abstraction and navigation processes. In addition, layout methods have been proposed which try to preserve similar parts in the heterogenous networks such as the visualization of two or three heterogenous networks in parallel planes in three dimensions is discussed in [21]. All these tools and approaches have in common that they provide solutions for analyzing a set of heterogeneous multivariate networks, but *not* at multiple levels.

In the following, we provide a short overview of techniques and ideas that might partly solve the problem of representing a set of heterogeneous networks distributed in several levels. All figures refer back to the sample three-level networks in Fig. 9.2, with green indicating the lowest level, red indicating the middle level, and blue indicating the upper level.

9.3.1 Approaches for Networks at Multiple Levels

Stacking

The most obvious visualization metaphor for networks on different levels is stacking. All networks on the same level are laid out (by using any more or less smart graph drawing algorithm) on a 2D plane, and then these planes are stacked in 3D (cf. Fig. 9.8). Multivariate data attached to the nodes or edges might be displayed within the planes themselves, as additional layers below or above the individual 2D planes, or separated into multiple coordinated views [59]. One of the existing example tools is VisLink [12], which is a general approach to show relationships between visualizations. In our special case, networks on levels are displayed on multiple 2D planes that can be arranged in the third dimension in various ways (in parallel, book-like, etc.). Relationships are represented as links—that is, as inter-plane edges. Here also, multivariate data can be represented inside of the planes or on additional individual planes. In the latter case, inter-plane edges might be used to point to the attached multivariate data. Note that—according to our definitions—heterogeneous networks are usually placed by VisLink to different planes and not on one plane. The most obvious drawback of such 3D techniques is their low level of scalability as well as clutter and perspective distortions especially when showing multivariate data in combination with the networks themselves.

Nesting

Another thinkable visualization approach is to use nesting for the explicit encoding of inter-level edges. This requires that mappings across consecutive levels are of $1 : n$ type. Figure 9.9 shows an example of how such an approach might look. Multivariate data could be represented as additional graphical features of the nested boxes/circles or within separated, coordinated views. Advantages are the "flat" layout which might support finding answers for specific tasks, such as the analysis of the aggregation results between co-author networks and institute collaboration networks (cf. Fig. 9.6). Another benefit is the integration of various interaction techniques similar to Treemaps [28]. Disadvantages are the visual complexity of the approach—induced by the mixture of link and box elements—as well as the possibly high space consumption. Although this approach can be used for the hierarchical

198   9 Heterogeneous Networks on Multiple Levels

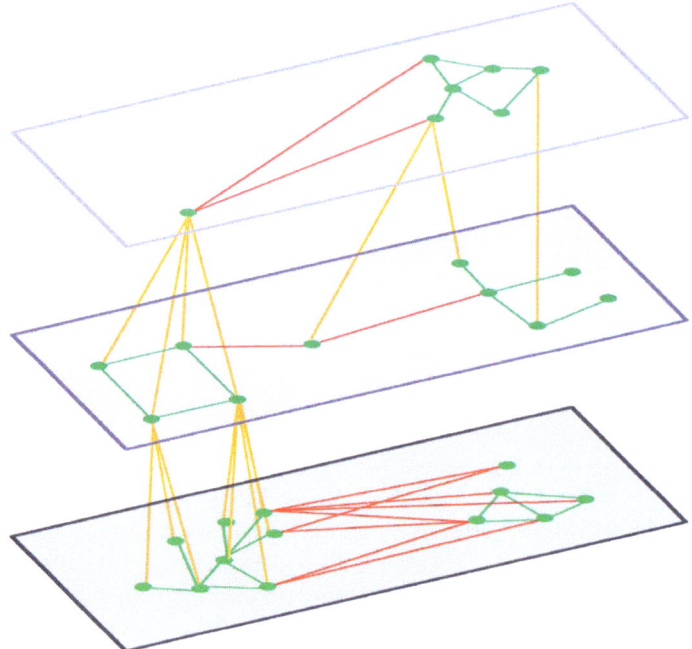

Fig. 9.8. Stacked 2D planes of network drawings which show the same networks as given in Fig. 9.2

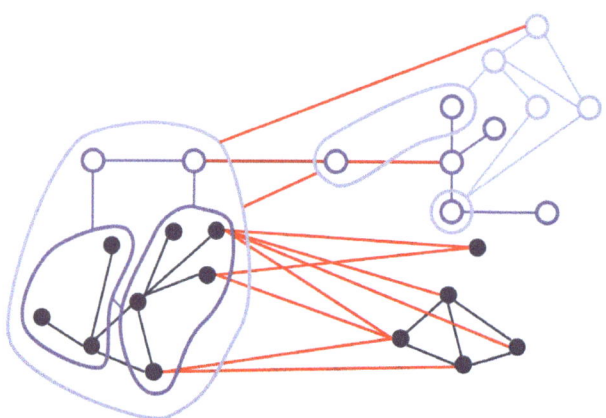

Fig. 9.9. Sketch of a nesting approach which shows the same networks as given in Fig. 9.2. Closed contours (incl. circles for nodes in level i that have no links to level $i-1$) are used to represent the $1:n$ mapping between levels.

presentation of one single network (e. g., clustering) [15], we are not aware of related works in the visualization literature for heterogeneous networks at multiple levels.

Alignment

The next approach uses one view per level which are all aligned to each other (cf. Fig 9.10). If a user brings a network part at one specific level into his view by zooming and panning, the related nodes in the other networks are shown simultaneously within the remaining coordinated views. Brushing can be used to selected individual nodes together with their neighbors at other levels as exemplified in the figure. Advantages of this idea are the simple metaphor which can be implemented easily as well as the rich interaction possibilities. Negative aspects might be the large space consumption of the many views and the missing inter-level edges. However, so-called context-preserving visual links [68] could solve this issue. Multivariate data could be displayed within the views (e. g., by glyphs or similar) or within additional coordinated views. The Entourage tool [49] realizes a similar approach with a special focus on contextual subsets, but without explicit encoding of the different levels. Here, contextually relevant pathways are displayed side-by-side together with a focus pathway, and only important parts of those context pathways are visible depending on the current selection in the focus pathway.

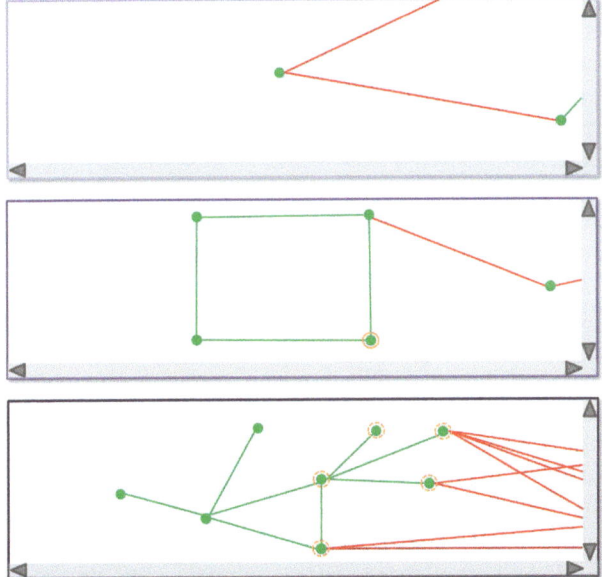

Fig. 9.10. Sketch of the alignment approach which shows the same networks as given in Fig. 9.2. Here, one node in the middle level was selected (orange halo); its neighbors in the lower level were highlighted accordingly (dashed orange halo).

9.3.2 Challenges and Future Directions

Heterogeneous networks on multiple levels are not easy to visualize even without attached multivariate data. One reason is the sheer size of the involved networks. Most graph drawing methods do not scale well. Another reason is the specific structure that is inherently given by the set of heterogeneous networks and the levels themselves: it is not efficiently reflected by most visualization approaches. Clever interaction techniques might help here, but in contrast to the visual analysis of single networks (or perhaps also heterogeneous networks within one level) we do not have a good understanding which interaction techniques and analytical methods work best in this context. More work has to be done to develop new visual representations and interaction metaphors to solve the specific problems and tasks described in the previous sections. This process has to be accompanied by user studies. Performing good and reliable evaluations is a challenge on its own, and we refer to the book [57] for further reading.

In this chapter, we restricted ourselves to $1 : n$ mappings across consecutive levels which can be sufficiently motivated by many concrete data sets and analysis tasks in our applications fields. However, there are—of course—situations in practice that demand universal $n : m$ relationships between network elements in different levels that must not be consecutive. Section 9.2.2 briefly exemplifies this issue. Because of the structural flexibility which comes with such general approaches, visualization experts have difficulties with the development of novel methods and tools that are able to handle those cases. From the perspective of visual analytics, more research has to be done in order to improve/facilitate the analysis processes. The analysis of heterogeneous networks on multiple levels is usually not done by just one analyst. Usually, several people work together—whether it be at one place (co-located) or several places (distributed), or whether it be at one specific period in time or at several different times. Our visualization and analysis tools should be able to support such collaborative work, record analysis sessions, support annotations by the users, and provide some guidance during the analysis process [37]—that is, a visualization should support "guided analytics to lead analysts through workflows for common tasks" [23].

For the integration of multivariate data into heterogeneous networks on multiple levels initial ideas have been proposed, but so far we have not really solved this problem. Both the network topologies and the attached multivariate data together are of great importance to the analysts, and in many tasks both is needed to solve specific questions and gain insights into the data. All ideas presented in Sect. 9.3.1 have the tendency to pay more regard to the network topology and not so much on the multivariate data. Such data can be added via glyphs or coordinated views, but this is not enough to discover patterns between the data and the network structure. Vice versa, if we focus on the multivariate data such as done by attribute-based network layouts [64, 71] and similar approaches [32], we run into the same problems.

Thus, finding an appropriate bunch of techniques for the common analysis of multivariate data within networks of networks is still an unsolved challenge.

Acknowledgments

We would like to thank all participants of the Dagstuhl Seminar ♯13201 [36] for the fruitful discussions and Todd Theriault for carefully proof-reading our chapter. This research is supported in part by the U.S. National Institutes of Health under Grant No. U01 GM098959, the German Ministry of Education and Research under Grant 0101-31P7126, and the German Academic Exchange Service (DAAD) under Grand 54391720.

References

1. Abuthawabeh, A., Beck, F., Zeckzer, D., Diehl, S.: Finding Structures in Multi-Type Code Couplings with Node-Link and Matrix Visualizations. In: Proceedings of the First IEEE Working Conference on Software Visualization, VISSOFT 2013 (2013)
2. Abuthawabeh, A., Zeckzer, D.: IMMV: An Interactive Multi-Matrix Visualization for Program Comprehension. In: Proceedings of the First IEEE Working Conference on Software Visualization, VISSOFT 2013 (2013)
3. Albrecht, M., Kerren, A., Klein, K., Kohlbacher, O., Mutzel, P., Paul, W., Schreiber, F., Wybrow, M.: On open problems in biological network visualization. In: Eppstein, D., Gansner, E.R. (eds.) GD 2009. LNCS, vol. 5849, pp. 256–267. Springer, Heidelberg (2010)
4. Bader, G.D., Cary, M.P., Sander, C.: Pathguide: a pathway resource list. Nucleic Acids Research 34, D504–D506 (2006)
5. Balasundaram, B., Butenko, S.: Network clustering. In: Junker, B.H., Schreiber, F. (eds.) Analysis of Biological Networks. Wiley Series on Bioinformatics, Computational Techniques and Engineering, pp. 113–138. Wiley (2008)
6. Bezerianos, A., Chevalier, F., Dragicevic, P., Elmqvist, N., Fekete, J.D.: Graphdice: A system for exploring multivariate social networks. Computer Graphics Forum (Proc. EuroVis 2010) 29(3), 863–872 (2010)
7. Borisjuk, L., Hajirezaei, M., Klukas, C., Rolletschek, H., Schreiber, F.: Integrating data from biological experiments into metabolic networks with the DBE information system. In Silico Biology 5, e11 (2004)
8. Byelas, H., Bondarev, E., Telea, A.: Visualization of areas of interest in component-based system architectures. In: Proceedings of the 32nd Euromicro Conference on Software Engineering and Advanced Applications, pp. 160–169. IEEE Computer Society Press (2006)
9. Börner, K., Sanyal, S., Vespignani, A.: Network science. In: Cronin, B. (ed.) Annual Review of Information Science and Technology, pp. 537–607. Information Today, Inc./American Society for Information Science and Technology (2007)
10. Christakis, N.A., Fowler, J.H.: The spread of obesity in a large social network over 32 years. New England Journal of Medicine 357, 370–379 (2007)
11. Cohen-Cole, E., Fletcher, J.M.: Detecting implausible social network effects in acne, height, and headaches: longitudinal analysis. BMJ, 337 (2008)

12. Collins, C., Carpendale, S.: Vislink: Revealing relationships amongst visualizations. IEEE Transactions on Visualization and Computer Graphics 13(6), 1192–1199 (2007)
13. Czauderna, T., Klukas, C., Schreiber, F.: Editing, validating and translating of SBGN maps. Bioinformatics 26(18), 2340–2341 (2010)
14. Demir, E., Cary, M.P., Paley, S., Fukuda, K., Lemer, C., Vastrik, I., Wu, G., D'Eustachio, P., Schaefer, C., Luciano, J., Schacherer, F., Martinez-Flores, I., Hu, Z., Jimenez-Jacinto, V., Joshi-Tope, G., Kandasamy, K., Lopez-Fuentes, A.C., Mi, H., Pichler, E., Rodchenkov, I., Splendiani, A., Tkachev, S., Zucker, J., Gopinath, G., Rajasimha, H., Ramakrishnan, R., Shah, I., Syed, M., Anwar, N., Babur, O., Blinov, M., Brauner, E., Corwin, D., Donaldson, S., Gibbons, F., Goldberg, R., Hornbeck, P., Luna, A., Murray-Rust, P., Neumann, E., Ruebenacker, O., Reubenacker, O., Samwald, M., van Iersel, M., Wimalaratne, S., Allen, K., Braun, B., Whirl-Carrillo, M., Cheung, K.H., Dahlquist, K., Finney, A., Gillespie, M., Glass, E., Gong, L., Haw, R., Honig, M., Hubaut, O., Kane, D., Krupa, S., Kutmon, M., Leonard, J., Marks, D., Merberg, D., Petri, V., Pico, A., Ravenscroft, D., Ren, L., Shah, N., Sunshine, M., Tang, R., Whaley, R., Letovksy, S., Buetow, K.H., Rzhetsky, A., Schachter, V., Sobral, B.S., Dogrusoz, U., McWeeney, S., Aladjem, M., Birney, E., Collado-Vides, J., Goto, S., Hucka, M., Novere, N.L., Maltsev, N., Pandey, A., Thomas, P., Wingender, E., Karp, P.D., Sander, C., Bader, G.D.: The BioPAX community standard for pathway data sharing. Nature Biotechnology 28(9), 935–942 (2010)
15. Dogrusoz, U., Giral, E., Cetintas, A., Civril, A., Demir, E.: A compound graph layout algorithm for biological pathways. In: Pach, J. (ed.) GD 2004. LNCS, vol. 3383, pp. 442–447. Springer, Heidelberg (2005)
16. Dunne, C., Henry-Riche, N., Lee, B., Metoyer, R., Robertson, G.: Graphtrail: analyzing large multivariate, heterogeneous networks while supporting exploration history. In: Proceedings of the SIGCHI Conference on Human Factors in Computing Systems, CHI 2012, pp. 1663–1672. ACM, New York (2012), http://doi.acm.org/10.1145/2207676.2208293
17. Dwyer, T., Hong, S.H., Koschützki, D., Schreiber, F., Xu, K.: Visual analysis of network centralities. In: Misue, K., Sugiyama, K., Tanaka, J. (eds.) Proc. Asia-Pacific Symposium on Information Visualization (APVis 2006). CRPIT, vol. 60, pp. 189–198. ACS (2006)
18. Fernández-Suárez, X.M., Galperin, M.Y.: The 2013 Nucleic Acids Research database issue and the online molecular biology database collection. Nucleic Acids Research 41, D1–D7 (2013)
19. Freire, M., Plaisant, C., Shneiderman, B., Golbeck, J.: Manynets: an interface for multiple network analysis and visualization. In: Proceedings of the SIGCHI Conference on Human Factors in Computing Systems, CHI 2010, pp. 213–222. ACM, New York (2010), http://doi.acm.org/10.1145/1753326.1753358
20. Funahashi, A., Matsuoka, Y., Jouraku, A., Kitano, H., Kikuchi, N.: CellDesigner: a modeling tool for biochemical networks. In: Proceedings of the 38th Conference on Winter Simulation, pp. 1707–1712. Winter Simulation Conference (2006)
21. Fung, D.C.Y., Hong, S.H., Koschützki, D., Schreiber, F., Xu, K.: Visual analysis of overlapping biological networks. In: Proceedings of the 13th International Conference on Information Visualisation, IV 2009, pp. 337–342. IEEE Computer Society Press (2009)

22. Gehlenborg, N., O'Donoghue, S.I., Baliga, N.S., Goesmann, A., Hibbs, M.A., Kitano, H., Kohlbacher, O., Neuweger, H., Schneider, R., Tenenbaum, D., Gavin, A.C.: Visualization of omics data for systems biology. Nature Methods 7, S56–S68 (2010)
23. Heer, J., Shneiderman, B.: Interactive dynamics for visual analysis. Communication of the ACM 55(4), 45–54 (2012), http://doi.acm.org/10.1145/2133806.2133821
24. Herman, I., Melançon, G., Marshall, M.S.: Graph visualization and navigation in information visualization: A survey. IEEE Transactions on Visualization and Computer Graphics 6(1), 24–43 (2000)
25. Hu, Z., Hung, J.H., Wang, Y., Chang, Y.C., Huang, C.L., Huyck, M., DeLisi, C.: VisANT 3.5: Multi-scale network visualization, analysis and inference based on the gene ontology. Nucleic Acids Research 37(Web Server issue), W115–W121 (2009)
26. Hucka, M., Finney, A., Sauro, H.M., Bolouri, H., Doyle, J.C., Kitano, H., Arkin, A.P., Bornstein, B.J., Bray, D., Cornish-Bowden, A., Cuellar, A.A., Dronov, S., Gilles, E.D., Ginkel, M., Gor, V., Goryanin, I., Hedley, W.J., Hodgman, T.C., Hofmeyr, J.H., Hunter, P.J., Juty, N.S., Kasberger, J.L., Kremling, A., Kummer, U., Novere, N.L., Loew, L.M., Lucio, D., Mendes, P., Minch, E., Mjolsness, E.D., Nakayama, Y., Nelson, M.R., Nielsen, P.F., Sakurada, T., Schaff, J.C., Shapiro, B.E., Shimizu, T.S., Spence, H.D., Stelling, J., Takahashi, K., Tomita, M., Wagner, J., Wang, J.: The systems biology markup language (SBML): a medium for representation and exchange of biochemical network models. Bioinformatics 19, 524–531 (2003)
27. van Iersel, M.P., Kelder, T., Pico, A.R., Hanspers, K., Coort, S., Conklin, B.R., Evelo, C.: Presenting and exploring biological pathways with PathVisio. BMC Bioinformatics 9, 399.1–399.9 (2008)
28. Johnson, B., Shneiderman, B.: Tree-maps: a space-filling approach to the visualization of hierarchical information structures. In: Proceedings of the 2nd Conference on Visualization (Vis 1991), pp. 284–291. IEEE Computer Society Press, Los Alamitos (1991), http://portal.acm.org/citation.cfm?id=949607.949654
29. Junker, A., Rohn, H., Schreiber, F.: Visual analysis of transcriptome data in the context of anatomical structures and biological networks. Frontiers in Plant Science 3, 252 (2012)
30. Junker, B.H., Klukas, C., Schreiber, F.: VANTED: A system for advanced data analysis and visualization in the context of biological networks. BMC Bioinformatics 7, 109 (2006)
31. Jusufi, I.: Multivariate Networks: Visualization and Interaction Techniques. Ph.D. Thesis, Linnaeus University, Växjö, Sweden (2013)
32. Jusufi, I., Kerren, A., Zimmer, B.: Multivariate network exploration with JauntyNets. In: Proceedings of the 17th International Conference on Information Visualisation (IV 2013), pp. 19–27. IEEE Computer Society Press (2013)
33. Juty, N., Le Novère, N., Laibe, C.: Identifiers.org and MIRIAM registry: community resources to provide persistent identification. Nucleic Acids Research 40(1), D580–D5869 (2012)
34. Kanehisa, M., Goto, S.: KEGG: Kyoto encyclopedia of genes and genomes. Nucleic Acids Research 28(1), 27–30 (2000)

35. Kerren, A., Köstinger, H., Zimmer, B.: Vincent – visualisation of network centralities. In: Proceedings of the International Conference on Information Visualization Theory and Applications (IVAPP 2012), pp. 703–712. INSTICC (2012)
36. Kerren, A., Purchase, H., Ward, M.O.: Information Visualization – Towards Multivariate Network Visualization (Dagstuhl Seminar 13201). Dagstuhl Reports 3(5), 19–42 (2013),
http://drops.dagstuhl.de/opus/volltexte/2013/4177
37. Kerren, A., Schreiber, F.: Toward the role of interaction in visual analytics. In: Proceedings of the Winter Simulation Conference, WSC 2012, pp. 420:1–420:13. Winter Simulation Conference (2012),
http://dl.acm.org/citation.cfm?id=2429759.2430303
38. Kerren, A., Stasko, J.T., Fekete, J.-D., North, C. (eds.): Information Visualization. LNCS, vol. 4950, pp. 65–91. Springer, Heidelberg (2008)
39. Klukas, C., Schreiber, F.: Integration of -omics data and networks for biomedical research with Vanted. Journal of Integrative Bioinformatics 7(2), 112 (2010)
40. Knodel, J., Muthig, D., Naab, M.: Understanding software architectures by visualization–an experiment with graphical elements. In: Working Conference on Reverse Engineering, pp. 39–50 (2006)
41. Köhler, J., Baumbach, J., Taubert, J., Specht, M., Skusa, A., Rüegg, A., Rawlings, C., Verrier, P., Philippi, S.: Graph-based analysis and visualization of experimental results with ONDEX. Bioinformatics 22(11), 1383–1390 (2006)
42. Kolpakov, F.A.: BioUML – framework for visual modeling and simulation of biological systems. In: Proceedings of the International Conference on Bioinformatics of Genome Regulation and Structure, pp. 130–133. Springer (2002)
43. Kono, N., Arakawa, K., Ogawa, R., Kido, N., Oshita, K., Ikegami, K., Tamaki, S., Tomit, M.: Pathway Projector: Web-based zoomable pathway browser using KEGG atlas and Google maps API. PLoS ONE 4(11), e7710 (2009)
44. Koschützki, D.: Network centralities. In: Junker, B.H., Schreiber, F. (eds.) Analysis of Biological Networks. Wiley Series on Bioinformatics, Computational Techniques and Engineering, pp. 65–84. Wiley (2008)
45. Küntzer, J., Backes, C., Blum, T., Gerasch, A., Kaufmann, M., Kohlbacher, O., Lenhof, H.P.: Bndb - the biochemical network database. BMC Bioinformatics 8, 367 (2007)
46. von Landesberger, T., Kuijper, A., Schreck, T., Kohlhammer, J., van Wijk, J., Fekete, J.D., Fellner, D.: Visual analysis of large graphs: State-of-the-art and future research challenges. Computer Graphics Forum 30(6), 1719–1749 (2011),
http://dx.doi.org/10.1111/j.1467-8659.2011.01898.x
47. Le Novère, N., Finney, A., Hucka, M., Bhalla, U.S., Campagne, F., Collado-Vides, J., Crampin, E.J., Halstead, M., Klipp, E., Mendes, P., Nielsen, P., Sauro, H., Shapiro, B., Snoep, J.L., Spence, H.D., Wanner, B.L.: Minimum information requested in the annotation of biochemical models (MIRIAM). Nature Biotechnology 23(12), 1509–1515 (2005)
48. Le Novère, N., Hucka, M., Mi, H., Moodie, S., Schreiber, F., Sorokin, A., Demir, E., Wegner, K., Aladjem, M.I., Wimalaratne, S.M., Bergman, F.T., Gauges, R., Ghazal, P., Kawaji, H., Li, L., Matsuoka, Y., Villéger, A., Boyd, S.E., Calzone, L., Courtot, M., Dogrusoz, U., Freeman, T.C., Funahashi, A., Ghosh, S., Jouraku, A., Kim, S., Kolpakov, F., Luna, A., Sahle, S., Schmidt, E., Watterson, S., Wu, G., Goryanin, I., Kell, D.B., Sander, C., Sauro, H., Snoep, J.L., Kohn, K., Kitano, H.: The Systems Biology Graphical Notation. Nature Biotechnology 27(8), 735–741 (2009)

49. Lex, A., Partl, C., Kalkofen, D., Streit, M., Wasserman, A.M., Gratzl, S., Schmalstieg, D., Pfister, H.: Entourage: Visualizing relationships between biological pathways using contextual subsets. IEEE Transactions on Visualization and Computer Graphics (InfoVis 2013) 19(12), 2536–2545 (2013)
50. Mehlhorn, H., Schreiber, F.: TransID – the flexible identifier mapping service. In: Proc. International Symposium on Integrative Bioinformatics, pp. 112–121 (2012)
51. Mi, H., Schreiber, F., Novère, N.L., Moodie, S., Sorokin, A.: Systems biology graphical notation: Activity flow language level1. In: Nature Precedings (2009)
52. Milo, R., Shen-Orr, S., Itzkovitz, S., Kashtan, N., Chklovskii, D., Alon, U.: Network motifs: Simple building blocks of complex networks. Science 298(5594), 824–827 (2002)
53. Moodie, S., Novère, N.L., Sorokin, A., Mi, H., Schreiber, F.: Systems biology graphical notation: Process description language level 1. In: Nature Precedings (2009)
54. Mueller, L.A., Zhang, P., Rhee, S.Y.: AraCyc: a biochemical pathway database for Arabidopsis. Plant Physiology 132(2), 453–460 (2003)
55. Novère, N.L., Moodie, S., Sorokin, A., Schreiber, F., Mi, H.: Systems biology graphical notation: Entity relationship language level 1. In: Nature Precedings (2009)
56. Partl, C., Kalkofen, D., Lex, A., Kashofer, K., Streit, M., Schmalstieg, D.: enroute: Dynamic path extraction from biological pathway maps for in-depth experimental data analysis. In: Proceedings of the 2012 IEEE Symposium on Biological Data Visualization (BioVis) BIOVIS 2012, pp. 107–114. IEEE Computer Society, Washington, DC (2012), http://dx.doi.org/10.1109/BioVis.2012.6378600
57. Purchase, H.: Experimental Human-Computer Interaction: A Practical Guide With Visual Examples. Cambridge University Press, New York (2012), http://eprints.gla.ac.uk/78680/
58. Rao, R., Card, S.K.: The table lens: merging graphical and symbolic representations in an interactive focus+context visualization for tabular information. In: CHI 1994: Conference Companion on Human Factors in Computing Systems, p. 222. ACM (1994)
59. Roberts, J.C.: Exploratory visualization with multiple linked views. In: MacEachren, A., Kraak, M.J., Dykes, J. (eds.) Exploring Geovisualization. Elseviers (2004), http://www.cs.kent.ac.uk/pubs/2004/1822
60. Rohn, H., Junker, A., Hartmann, A., Grafahrend-Belau, E., Treutler, H., Klapperstück, M., Czauderna, T., Klukas, C., Schreiber, F.: VANTED v2: a framework for systems biology applications. BMC Systems Biology 6(139) (2012)
61. Schreiber, F., Colmsee, C., Czauderna, T., Grafahrend-Belau, E., Hartmann, A., Junker, A., Junker, B.H., Klapperstück, M., Scholz, U., Weise, S.: MetaCrop 2.0: managing and exploring information about crop plant metabolism. Nucleic Acids Research 40(1), D1173–D1177 (2012)
62. Shen, Z., Ma, K.L.: Mobivis: A visualization system for exploring mobile data. In: Proceedings of IEEE Pacific Visualization Symposium, pp. 175–182. IEEE VGTC (2008)
63. Shen, Z., Ma, K.L., Eliassi-Rad, T.: Visual analysis of large heterogeneous social networks by semantic and structural abstraction. IEEE Transactions on Visualization and Computer Graphics 12(6), 1427–1439 (2006), http://dx.doi.org/10.1109/TVCG.2006.107

64. Shneiderman, B., Aris, A.: Network visualization by semantic substrates. IEEE Transaction on Visualization and Computer Graphics 12(5) (2006)
65. Smoot, M.E., Ono, K., Ruscheinski, J., Wang, P.L., Ideker, T.: Cytoscape 2.8: new features for data integration and network visualization. Bioinformatics 27(3), 431–432 (2011)
66. Sommer, B., Künsemöller, J., Sand, N., Husemann, A., Rumming, M., Kormeier, B.: Cellmicrocosmos 4.1 - an interactive approach to integrating spatially localized metabolic networks into a virtual 3d cell environment. In: Fred, A.L.N., Filipe, J., Gamboa, H. (eds.) Proc. International Conference on Bioinformatics, pp. 90–95 (2010)
67. Stasko, J., Görg, C., Liu, Z.: Jigsaw: supporting investigative analysis through interactive visualization. Information Visualization 7(2), 118–132 (2008), http://dx.doi.org/10.1145/1466620.1466622
68. Steinberger, M., Waldner, M., Streit, M., Lex, A., Schmalstieg, D.: Context-preserving visual links. IEEE Transactions on Visualization and Computer Graphics (InfoVis 2011) 17(12), 2249–2258 (2011)
69. Ward, M., Grinstein, G., Keim, D.A.: Interactive Data Visualization: Foundations, Techniques, and Application. A.K. Peters, Ltd. (2010)
70. Ware, C.: Information Visualization: Perception for Design, 2nd edn. Morgan Kaufmann (2004)
71. Wattenberg, M.: Visual exploration of multivariate graphs. In: Proceedings of the SIGCHI Conference on Human Factors in Computing Systems (CHI 2006), pp. 811–819. ACM, New York (2006)
72. Zimmer, B., Jusufi, I., Kerren, A.: Analyzing multiple network centralities with ViNCent. In: Proceedings of SIGRAD 2012: Interactive Visual Analysis of Data, Växjö, Sweden, November 29-30. Linköping Electronic Conference Proceedings, vol. 81, pp. 87–90. Linköping University Electronic Press (2012)

10
Scalability Considerations for Multivariate Graph Visualization

T.J. Jankun-Kelly, Tim Dwyer, Danny Holten, Christophe Hurter, Martin Nöllenburg, Chris Weaver, and Kai Xu

Scalability in visualization is a challenge: How do we choose to show more items than can be easily rendered upon a screen or understood by a human effectively? Multivariate graph visualization adds additional wrinkles in that nodes and edges are no longer atomic entities. Rather, they are repositories for further rich information. In information seeking, the mantra attributed to Ben Shneiderman succinctly outlines a path to visual scalability: "Overview first, zoom, then details-on-demand" [69]. While this is good guidance, naively presenting the whole universe of data as an initial "overview", leads to dense, unreadable displays (Fig. 10.1). To provide insightful visualizations at large scale for multivariate graphs, we must understand what our visual, cognitive, and architectural limits are, then explore approaches to mitigate these limitations. Detailed views must offer useful affordances for navigation to other views. The goals of this chapter are to identify the challenges and the state-of-the-art in these areas.

At large scale, dense multivariate graphs devolve into *hairballs* (dense collections of nodes with heavily over plotted edges) or *snowy wastes* (highly populated matrix diagrams with visually random structure) if the entire structure is shown (Fig. 10.1). Perceptual and cognitive psychology outline what human visual and mental limitations interfere with understanding such dense views; additionally, there are hardware factors which band the amount of graph data that can be rendered in a timely matter. By understanding these limitations, outlined in Sect. 10.1, designers can utilize the strategies explored in Sect. 10.2 to show only what is needed when it is needed. Use of these strategies, and further studies on the limits of scalability, are also presented in Sect. 10.3 as a means to guide further research. We conclude with a summary of challenges in scalable, multivariate graph visualization.

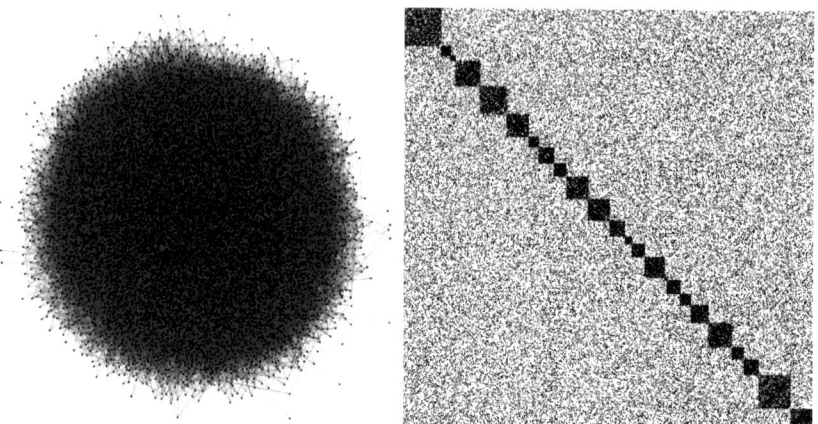

Fig. 10.1. Large (especially scale-free) graphs turn in to *hairballs* which make understanding structure difficult (left). Matrix views can help, but have limited ability to convey transitive structure at multiple levels and devolve into *snowy wastes* (right).

10.1 Limits of Visualization

When attempting to display and understand data, there are limits to what the can be humanly perceived and understood and what can be computed and displayed. To work around these limitations, they must be first understood. In this section, we examine both perceptual and computational/hardware based limitations to set the stage for the larger discussion of scalability in the next sections.

10.1.1 Limits of Visual Acuity

As a sensor, the eye has several different resolving powers or *acuities*. These acuities are measured based upon *visual angle*, the angle the viewed object(s) subtends with respect to the eye. Ophthalmologists recognize four main acuities: detection acuity (the smallest size an object can be before it cannot be seen), recognition acuity (smallest size at which an object can be identified), resolution acuity (smallest distance between two objects before they seem to merge), and localization acuity (smallest amount of visual change that can be measured between two visible objects) [87]. Perception literature in visualization has focused on the latter two, especially on point acuity (resolvability of two adjacent points), stereo acuity (the ability to resolve objects at depth), and vernier acuity (the ability to determine if two line segments are collinear) [76] (Fig. 10.2); point acuity is a form of resolution (or ordinary) acuity and stereo and vernier are localization acuities (*hyperacuities*). Resolution/point acuity are the primary acuities from the

10.1 Limits of Visualization

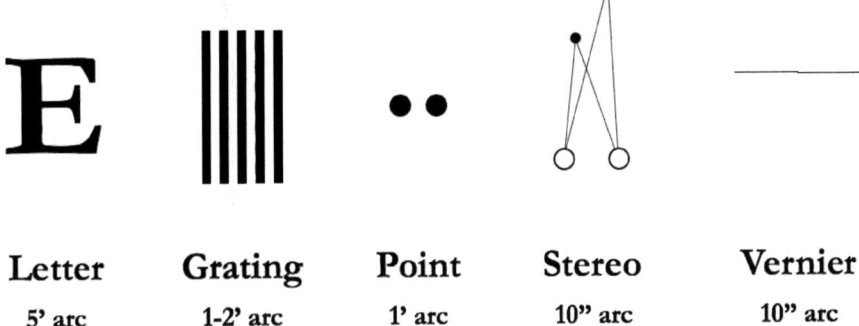

Fig. 10.2. Important acuities in visualization (after Ware [76])

standpoint of graph visualization design—it is important that two separate edges, nodes, or matrix elements be resolvable.

Assuming perfect vision, standard point acuity is one arc minute per cycle—i.e., two lines are perceived as distinct when one arc minute separates them.[1] Thus, roughly two pixels per arc minute would generate a maximally point resolvable display.[2] Though this pixel density will not be able to perfectly resolve hyperacuities, which are resolvable to 10 arc seconds, antialiasing can be used to effectively resolve hyperacuities to sub-pixel resolution [76]. At a viewing distance of 57 cm, this pixel density corresponds to 121 ppcm (pixels per centimeter); at reading distance, the density is 229 ppcm. If we relax this constraint to allow point acuity sufficient for legal driving in most countries, about half that of perfect acuity [55, 74], the pixel density would need to be one pixel per arc minute, or about 60 ppcm and 118 ppcm at viewing and reading distances respectively. For comparison, the Retina displays from Apple vary from 128 ppcm on the iPhone 5 (generally used closer to reading distance) to 87 for the Retina display laptops [88].

Given these minimum requirements for perceptual resolvability, node density (the number of nodes on the screen per area) cannot exceed roughly 1 per 2 pixels if 1 pixel representations are to be used for maximum discrimination; note, this limits our ability to indicate multivariate nodes/edges to using only the visual variables of hue and luminance. Thus, to use the MacBook Retina display as an exemplar, graphs exceeding roughly 2 million nodes would strain resolvability assuming the entire display was used and no connectivity information was displayed. If a node-like representation is used, however, this number drops to 1 million as edges must connect elements already on display. For matrix diagrams, the maximum limit is about 0.5

---

[1] "Perfect acuity" here is taken as an accepted average; this acuity will vary over a population.

[2] Two pixels per minute are needed to satisfy sampling theory—if we want to visually detect one pixel per arc minute, we need twice that many to satisfy the Nyquist criterion [24].

million nodes so neighboring connections are resolvable; if this is not desired, a packed representation still only represents about 1.4 million nodes as the matrix diagram's symmetric display scales quadratically with nodes. These patterns are summarized in Fig. 10.3. Exceeding these numbers means that the individual elements of the graph cannot be separated. However, staying below these numbers is not sufficient for comprehension of the graph—though elements may be perceivable, they may still exhaust cognitive resources as discussed next.

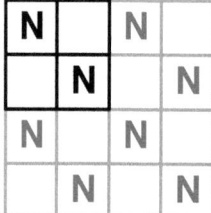 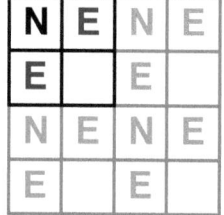 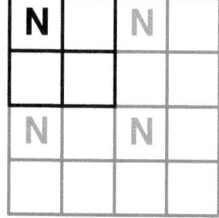

Fig. 10.3. Maximum descriminability for unconnected graphs (left), node-link diagrams (middle), and matrix diagrams (right). Each box is a pixel.

10.1.2 Cognitive Limits

Perception can be thought of grossly as a two stage process where elements are first imaged by the eye-sensor and then understood by the brain. This latter, cognitive step of perception has limitations not tied to the processing power of the eye discussed previously. Instead of focusing on raw node/element density as previously discussed, this section focuses more upon what limits of combined elements such as hue and luminance can be used, especially when used conjunctively as is often the case in multivariate visualization.

The cognitive stage of perception can be modeled as a hybrid bottom-up/top-down approach [75, 89, 90]. In the bottom-up stage, early vision separates perceived elements into feature maps (e.g., hue, orientation, length) with varying level of granularity. Elements in these maps are then compared to neighbors to measure differences. The user-driven top-down stage then searches these maps using the differences to find features of interest. As an example, if node luminance in a graph matrix display encodes weight, differences in luminance would be used as the perceptual search criteria. Thus, the levels of difference that can be encoded in the feature maps limit these lower level cognitive comparisons.

Ward et al. [75] summarize the absolute discriminability of perceptual features relevant to visualization, based upon work by Miller [54]. For item size, four to five different levels are accurately gauged; 10 levels for hue and 5 for luminance; and roughly 7 levels for line length and 8 for orientation. These levels are not independent when combined for multi-feature encoding (a popular approach in multivariate graph visualization)—only about 12 different combinations of hue/luminance can be separated (as opposed to 40–50 if they were independent) and roughly 17 levels for combinations of hue, luminance, and size. Thus, the number of perceptual values that can be used for absolute judgements is very limited—hairballs and snowy wastes will quickly exceed these capacities.

As cognition happens at multiple levels, studies have also investigated how graph features such as paths are comprehended at latter stages in cognition. These studies provide guidance on what approaches are suitable for multivariate graph visualization. A common theme of these is the limitation of what is displayed simultaneously. A variety of studies have investigate visual search of graphs, either of specific properties such as shortest paths or for more general search. In node-link representations, path comprehension suffers when edge crossings occur [60, 61], especially when dealing with crossings over shortest paths [79]. Limiting such crossing require reducing edges or novel layouts, both approaches discussed later. Using 2.5D displays can also help graph structure comprehension when used appropriately in node-link diagrams [12, 78]; here, occlusion helps limits what is displayed. Increasing visual separation in combination with limited motion cues can also help visual search in graphs [77]. Structure can be perceived with matrix diagrams by removing edges, with more accurate reading at larger graph sizes [30].

10.1.3 Leveraging the Graphics Card (GPU)

In visualization, the human is only part of the process—computation is also required to generate the visualization. Just as there are limitations on the human, there are limitations on the computation. These can be limitations on the display technology, the graph data store, or on the means of computation on the graph. In this work, we focus on how graphics processors can enable scalable graph visualization and their limitations; for discussion on the limits and capabilities of displays and graph storage, we refer the reader to the relevant literature [2, 22].

This section describes how graphics cards can be used to address scalability issues in general and with respect to multivariate graph visualization (MGV) in particular. Sections 10.1.3 and 10.1.3 serve as an intro to and history of fixed-functionality and programmable graphics hardware as well as available programming APIs, respectively. This is followed by Sect. 10.1.3 describing typical tasks and/or application areas within graphics and (information) visualization related to MGV scalability issues. This section also gives real-world examples of GPU-based solutions designed to tackle MGV scalability issues from the perspective of rendering, interaction, and calculation.

GPU Pipeline—Fixed vs. Programmable

The rise of special-purpose graphics hardware for the accelerated monochrome and color display of 2D/3D raster and vector graphics began during the mid to late 1970s and widespread consumer adoption—especially of hardware 3D-acceleration solutions—was obtained during the late 1990s. Such hardware was originally built around "fixed functionality pipelines" (FFPs), i.e., special purpose hardware that supports a limited and fixed set of instructions (drawing commands) to display various types of graphics primitives. Typical FFPs support operations such as geometric transformation, lighting, and rasterization, all of which are necessary for displaying ("projecting") 3D graphics on 2D raster displays.

An example of early 2D/3D-accelerated FFPs are the processing pipelines on special-purpose hardware built by Silicon Graphics International (SGI) for use in their high-end graphics workstations during the early to mid 1990s. The late 1990s to early 2000s saw the widespread consumer adoption of more affordable mainstream 2D/3D-accelerated graphics FFPs (often used in game consoles) such as the Voodoo, early GeForce, and early Radeon graphics hardware by 3Dfx, NVIDIA, and ATI, respectively.

From the mid 2000s onwards, graphics hardware manufacturers as well as graphics API developers (see Section 10.1.3) gradually shifted their focus to programmable pipelines instead of FFPs. Programmable pipelines allow the graphics processing unit (GPU) to run proprietary code [58]. Such code can be used to implement new types of drawing commands and can even be used—although initially indirectly—to perform (non-graphics-related) computational tasks on a GPU, i.e., "general purpose computation on the GPU" or GPGPU [72]. The latter is useful because of the massive parallelism offered by GPUs as well as the ease with which GPUs generally handle vector and matrix operations; a direct result of the fact that 2D/3D transformations and projections within FFPs rely heavily on vector/matrix math.

GPU Programming—APIs and Pitfalls

Programming each level of the graphics card pipeline can be performed through different languages, such as NVidia's Cg, Microsoft's High-Level Shading Language (HLSL), and the OpenGL shading language (GLSL). Other specialized languages exist to do specific data processing: CUDA, OpenCL. If we exclude specific data processing languages (CUDA and OpenCL) which use specific data structures, output data must be stored in image textures. Graphics cards propose massive parallel computing but some pitfalls must be avoided in order to take advantage of this worthy power:

- Graphics card are optimized to compute data in parallel and therefore sequential algorithms cannot be paralyzed without insuring data integrity (memory protection). Reading and writing graphics memory is not possible

at the same time; this avoids memory corruption (one process reading at the same time another is updating the information). Synchronization features such as mutex or memory protection (atomic functions) much be avoided as much as possible. Specific computation technics can be applied such as MapReduce, a programming model for processing large data sets with a parallel, distributed algorithm on a cluster [35].
- Bottlenecks exist within the GPU processing, especially when transferring data between the CPU and the GPU. When this occurs, the graphics card needs to wait until every process has ended, and then start the memory transfer—a dramatically slower process. Memory transfer between the GPU and the CPU must be limited as much as possible.
- Many other pitfalls must be taken into account regarding every language, such as texture coordinates that differ between OpenGL and DirectX, debugging issues, and graphics card crashes that hinder the development process.

Multivariate Graph Visualisation (MGV) Scalability Issues

When dealing with large MGV, we identified three types of scalability limitations which are related to the InfoVis pipeline stages [6]: rendering, computation and interaction.

MGV can embed numerous items which need to be displayed. On the rendering stage, specific rendering techniques can be used to emphasize the rendering process and thus to improve data perception (an example is given in Fig. 10.4).

MGV can contain complex data structures. Layout algorithms, graph simplifications, data aggregations can be computed. Processing such information at the geometry level can be an issue when dealing with large MGV.

Finally, MGV can face scalability issues with interactive tools. Large dataset can hinder the data exploration process with system slowing down.

Instances of GPU Usages for MGV

Given the above, we identified the following GPU usages to address scalability issues with large MGV. The key to these techniques is how they overcome the limitations of the GPU mentioned previous to facilitate multivariate graph exploration—they use multi pass read-write cycles, minimize CPU-GPU memory transfer, and accommodate variation in graphical hardware:

Rendering: Graphics cards can render numerous items on the screen and thus can display large datasets. In the following examples, GPUs are used to display data and to perform image based rendering techniques. Auber developed Tulip [3], an information visualization framework dedicated to the analysis and visualization of relational data. This software uses GP-GPU techniques to render large multivariate graphs. McDonnel et al. [53] developed a framework and an application using shaders to display multivariate data based on the dataflow model with a final image based stage. In this final step, the multivariate data of the visualization are sampled in the resolution of the current view. A more specific rendering technique is used by Holten [39] to improve edge visualization by an interesting variation on standard alpha blending, i.e. how color transparency is combined. Sheepens et al. [67] used the GPU to compute density maps and then apply shading techniques to emphasize multivariate data on the density map of moving vessels.

Computation: Graphics cards can perform fast and parallel data processing, and be used to process information at the data level. Hurter et al. [46] use the GPU for interactive exploration of multivariate relational data. Given a spatial embedding of the data, in terms of a scatter plot or graph layout, the moleview uses a semantic lens which selects a specific spatial and attribute-related data range. The lens keeps the selected data in focus unchanged and continuously deforms the data out of the selection range in order to maintain the context around the focus. Animation is also performed between the bundled and the unbundled layout of a graph. Kernel Density Edge Bundling (KDEB) [44] computes bundled layouts of general graphs. For this, KDEB first transforms a given graph drawing into a density map using kernel density estimation. Next, it applies an image sharpening technique which progressively merges local height maxima by moving the convolved graph edges into the height gradient flow. This technique is also applied on dynamic graphs [45]. Graph bundling and the computation of its density has been investigated [51], and the GPU has been used directly for graph layout as well [26].

Interaction: Interaction is an important manipulation paradigm to perform data exploration. Graphics cards can be used to provide tools to help user to interact with large datasets. ScatterDice [17] helps the user to define the appropriated displayed variables with a smooth animation when changing visual configuration; GraphDice [5] uses the same paradigms but with graph. FromDaDy [47] uses related animation with GP-GPU techniques. In order to address dataset size issue, FromDaDy loads the whole dataset within the graphics card, so that when changing visual configuration, no memory transfer is needed. This helps to improve interaction with a fast and continuous animations. Furthermore, a GP-GPU technique is implemented to support brushing and data manipulation across multiple views. One can then brush trajectories, and with a pick and drop operation he or she can spread the brushed information across views. This interaction can be repeated to extract a set of relevant data, thus formulating complex queries. Each trajectory has a unique identifier. A texture (stored in the graphics card) contains the Boolean selection value of each trajectory. When the trajectory is brushed its value is set to true. The graphics card uses parallel rendering which prevents reading and writing in the same texture in a single pass. Therefore FromDaDy used

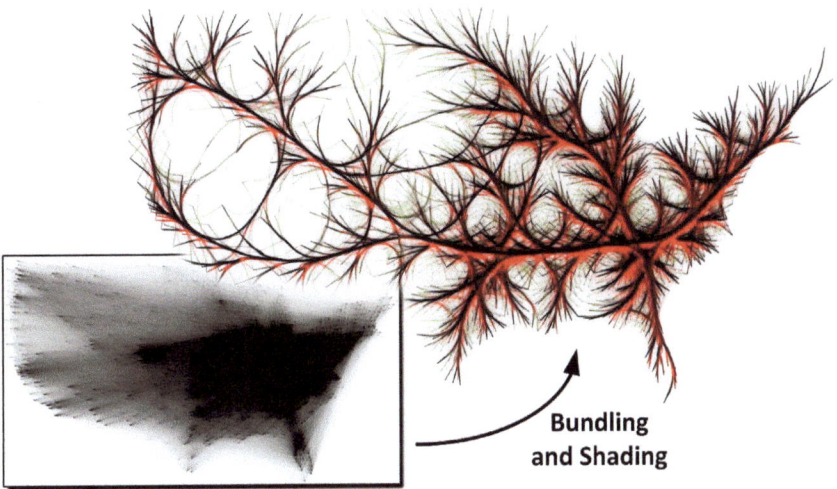

Fig. 10.4. County-to-county migration flow, (1091764 nodes, the Census 2000): people who moved between counties within 5 years. Original data only shows the outline of the USA (bottom), bundled [44] and shaded path (top) shows multiple information like East-West and north-South paths, shading shows data density.

a two-step rendering process: firstly it tests the intersection of the brushing shape and the point to be rendered to update the selected identifier texture, and, secondly, it draws all the points with their corresponding selected attribute (gray color if selected, visual configuration color otherwise).

10.2 Design Strategies for Scalable Multivariate Graph Visualization

The perceptual, cognitive, and technical factors presented in Sect. 10.1 limit the scalability of network visualization in general. In particular, we described the limitations of:

- visual acuity (10.1.1);
- human cognition (10.1.2);
- computer hardware (10.1.3); and
- computability of aesthetic and clear layout (10.1.2 and 10.1.3).

When faced with the increased amount and complexity of information that one typically encounters in multivariate networks, it is necessary to address scalability of visualization by additional means. For example, there may be rich tabular data associated with graph elements; graph elements may have myriad types; and graphs may be derived from underlying data in many different ways. In this section, we review various design strategies to support scalable interactive visualization of multivariate graphs, including very large ones. These strategies go beyond simply getting as much information onto the screen as possible. They also aim to make good use of available display real estate by transforming and reducing that information to facilitate exploration and analysis.

Chapter 6 describes interactive operations in terms of the information visualization reference model of Card et al. [7]. In the model, three transformation steps connect a progression of four modes of data representation from raw data (at one end) to displayed visuals (at the other end) (Fig. 6.1). In keeping with the reference model, we organize multivariate graph design strategies into the three following categories of transformations of information representations.

Data transformation and reduction strategies provide alternative *network compositions* by being selective about the type of structure and amount of information to show. These strategies use combinations of aggregation, projection, and filtering techniques to convert multivariate graph data sets into other data sets, particularly into alternative multivariate graphs having topologies and attributes that can be more readily and usefully displayed.

Visual mapping strategies provide alternative *network presentations* by mapping data dimensions and values into visual elements that efficiently communicate graph structure and multivariate attributes. The definition of *efficient* here depends on the application and the type of analysis being sought. These strategies often complement data transformation and reduction strategies by choosing mappings to suit the aggregated, projected, and filtered information of specific network perspectives.

View transformation strategies provide alternative *network perspectives* by providing a feedback loop for the analyst to interact with visual elements and the space in which they are shown. Visualizing multivariate networks in any real analysis application is not a static "batch" or "pipeline" process. View transformation strategies often complement other strategies by supporting not only navigation in view space and selection of data items, but also interactive changes to the functions and parameters used in data transformation and reduction and in visual mapping.

Weaver breaks down this progression of transformations into a more detailed model that specifically targets interactive visualization of complex multivariate data as networks. The model is implemented in Improvise [82] and has been used in a variety of graph visualization applications including to metavisualize multiple view coordination structure [83, 84] and to analyze individual differences in user categorization of starplot shapes [49] and geospatial relationships [50]. Figure 10.5 depicts the data transformation pipelines in this graph reference model, as customized for use in the Attribute Relationship Graph technique [86]. In this model, the three transformation steps in Card's model are expanded into three interdependent phases of transformation.

10.2 Design Strategies for Scalable Multivariate Graph Visualization

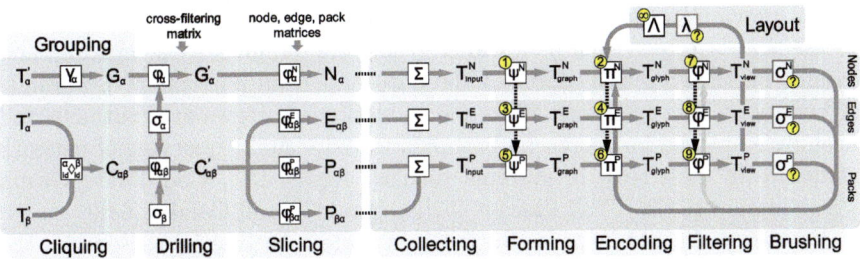

Fig. 10.5. The data transformation pipeline of Weaver [86] showing transformation of raw multivariate data into tractable graph views

Card's Data Transformation step expands into the first two phases: data projection and graph definition. The data projection phase (Fig. 10.5, left) aggregates occurrences (*Grouping*) and co-occurrences (*Cliquing*) of the data values in each dimension, then determines which occurrences and co-occurrence to show as a function of chosen dimensions (*Drilling*) and data values (*Slicing*). Graph definition consists of transformations (Fig. 10.5, center) that gather the aggregates into tables (*Collecting*) that are then mapped into a graph representation (*Forming*) consisting of primitive elements identified as nodes, edges and "packs". (Packs can be thought of as hyperedges that connect multiple nodes into semantically grouped aggregates, and are often shown as convex hulls around the connected nodes, as in Vizster [36]).

Card's Visual Mapping and View Transformation steps expand into a single graph visualization phase (Fig. 10.5, right). Parallel pipelines that take the sets of graph primitives generated in graph definition as input. The subsequent *Encoding*, *Filtering*, *Layout*, and *Brushing* transformations populate and support interaction with graph primitives in network, matrix, and other data views. Transformations are interdependent to capture the ways that one can expect graph elements to be coordinated in appearance and behavior, such as filtering of edges on whether their nodes are visible (as well as as a function of their co-occurrence data attributes), or reencoding edges when a node moves during automatic or interactive layout.

Together, the three strategies can be seen as axes in a rich design space. We use the graph reference model here as a frame of reference to discuss the work that has been done in this space. Below we explore each of the strategies in more detail, looking at how they are employed individually and in combination in various exemplar systems.

10.2.1 Data Transformation and Reduction

Visual bandwidth is finite, both on the production side (the graphics and display hardware), and on the consumption side (the perceptual and cognitive capabilities of the person trying to analyze the data); these are alluded to in the previous section. Data transformation and reduction techniques aim to use the available visual bandwidth to support foraging, sense-making, and insight by showing only parts of the data and from particular viewpoints. A variety of data transformation and reduction strategies are routinely and usefully employed for increasingly sophisticated visually querying of network structure in data sets, in systems such as Jigsaw [70], Coordinate Graph Visualization (CGV) [73], Ploceus [52], Orion [37], and Candid [68].

Whatever the data-size limitation in a certain setting is, as soon as the graph exceeds it, we can no longer show all information in a single static view. Instead, data reduction techniques must be applied to extract a task-specific neighbourhood of the larger graph for display. This extraction can be fully automatic, or semi-automatic according to the constraints of the user. As in Chapter 6, we are interested in data reduction operations necessary for producing limited views of very large graphs. In this chapter, however—with our focus on scalability—we review recent work that deals more specifically with the problem of extracting small tractable graph views from big tabular data. We distinguish three different data transformation and reduction approaches: aggregation, projection, filtering.

Aggregation Techniques

Graph aggregation techniques transform and reduce data sets by collecting data records into buckets in terms of commonalities shared by the raw or derived attributes of those records. The underlying principle of graph aggregation is, for a given graph $G = (V, E)$, to derive an aggregate graph $\widetilde{G} = (\widetilde{V}, \widetilde{E})$ with fewer vertices and edges. The goal is to compute $\widetilde{G}$ in such a way that it is a good coarse representation of G for the user's data analysis purposes. This data reduction process is also known as *granulation* [71].

In the graph reference model, graph aggregation happens in the grouping (for vertices) and cliquing (for edges) stage. These transformations perform unary and binary calculations to determine raw or derived attributes occurrences and co-occurrences, respectively. Using raw attribute values themselves is a basic approach but still highly useful for analysis; Jigsaw [70], Cross-Filtered Views [85], and Attribute Relationship Graphs [86] all support multivariate association an comparison tasks in this way.

Many systems provide several types of graph aggregation that entail more complex calculations of derived graph elements (i.e., with a one-to-many mapping from graph node or link to data). PivotGraph [81] uses roll-up for multivariate graphs to group nodes into equivalence classes based on their attribute values and create weighted edges as induced by the members of the different equivalence classes. Selection by restricting certain attribute values can be used to obtain the induced subgraphs. This technique originally does not focus on large graphs and depending on the attribute types and values, the number of equivalence classes may be too high. Orion [37] supports similar attribute-based aggregation of vertices for networks that are obtained in a previous step from relational database tables. In the Ploceus system [52], three types of aggregation-by-attribute are identified: *pivoting* is equivalent to PivotGraph's *roll-up* but specifically for categorical attributes; *binning* is used to describe grouping nodes by quantitative attributes divided into distinct ranges; *proximity grouping* is used to refer to more sophisticated clustering techniques involving distance functions of quantitative attributes.

More generally, one can perform arbitrary many-to-many calculations to generate derived data dimensions for grouping and cliquing. Clustering is a common approach, although one that requires great care to maintain responsiveness of the overall graph transformation pipeline. *Graph clustering* techniques partition the vertex set V into mutually disjoint clusters $C_1, \ldots, C_k$ with the objective that two vertices in the same vertex cluster C_i are sufficiently similar and two vertices from different clusters are sufficiently dissimilar. As we deal with multivariate graphs, there is a wide range of clustering methods that can be applied.

Graph-based clustering methods consider edges (possibly with weights) as an indicator for similarity and hence aim at finding a clustering with high intra-cluster edge density and low inter-cluster edge density. Fortunato's recent survey on community detection in graphs covers the state of the art in clustering algorithms [25]. On the other hand, attribute-based clustering methods are data mining techniques that consider each vertex as a point in a multi-dimensional space spanned by the multivariate vertex attributes. Using a (dis-)similarity measure defined in this space, clusters are derived based on this measure. Again, vertices in the same cluster should have high similarity, and vertices in different clusters should be dissimilar. Berkhin gives a recent review of the most common clustering methods in data mining [4]. Clustering of multivariate graphs ideally uses methods that combine connectivity information and attribute information in a configurable way, especially if attributes and edges are not highly correlated. Only few methods exist that take into account both types of information. Zhou et al. [94, 95] present a method to transform the attribute data of large graphs into additional graph edges and then apply a graph clustering algorithm to the augmented graph. Another combined method is DB-CSC by Günnemann et al. [31], which allows more flexible cluster shapes. Hadlak [33] describes clustering on time series behavior of time-varying attributes.

Of particular interest are hierarchical graph clustering methods [25, Chap. IV.B], where different clustering granularities can be represented between a single cluster containing everything at the top and singleton clusters at the bottom. In the graph reference model, hierarchies can be treated as multiple aggregations that are coincidentally related in terms of data type semantics. (In the graph visualization phase of the model, representation and interaction should reinforce these relationships.) Depending on the navigation strategy, different types of cuts or *frontiers* in the clustering tree can be applied. Thus it is possible to obtain rather uniform granularities for an aggregated overview graph or non-uniform granularities giving more details in a focus region and

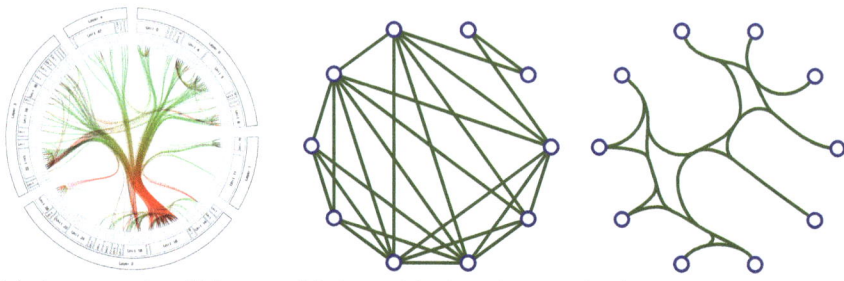

(a) An example of hierarchical edge bundling [39].   (b) A straight-line drawing (left) and a strict confluent drawing (right) of the same graph [20].

Fig. 10.6. Radial graph layouts using edge aggregation

less details in the further context. ASK-GraphView [1] is a system that applies hierarchical clustering for visualizing large graphs.

While the above methods are mostly concerned with aggregating vertices to reduce the graph size, there are also several techniques to aggregate edges. Since a graph of $|V|$ vertices can have $O(|V|^2)$ edges, graphs with relatively few vertices can already be too edge dense to be readable. Edge bundling methods [21, 27, 39, 40, 44, 45] aim to reduce visual clutter by visually grouping together edges between similar parts of the graph thus using fewer pixels to show the original set of edges. Visualizations using edge bundling are well suited to depict global connectivity patterns with reduced visual complexity. See Fig. 10.6(a) for an example. The topological information produced in the graph specification phase of the graph reference model can be used to aggregate edges, although having two or more (potentially interdependent) stages of aggregation complicates matters substantially; such complex interdependencies between the node and edge pipelines that define a graph visualization are beyond the scope of the graph reference model.

The concept of confluent graph drawing [10, 18–20, 43, 63], where two vertices are connected if and only if there is a smooth path between them, similarly merges and splits the curves representing edges in a visualization. But unlike edge bundling methods, confluent drawings are unambiguous or *information faithful* [57] since no false adjacencies are created. Confluent drawings can be used to display certain non-planar graphs without edge crossings. Figure 10.6(b) shows a straight-line and a confluent drawing of the same graph. Confluent drawing algorithms are not yet implemented in practical systems and some related decision problems are known to be NP-complete. Edge compression through Power Graph Analysis [13, 66] (see Fig. 10.7) is another technique for aggregating edges by replacing the edges of bipartite clique sub-graphs with single edges connecting the two sets of nodes in the bipartite clique. Power Graph compression is related to confluent drawing in that it also offers an unambiguous, information faithful representation of the original dense graph however practical techniques exist for their generation [16].

10.2 Design Strategies for Scalable Multivariate Graph Visualization

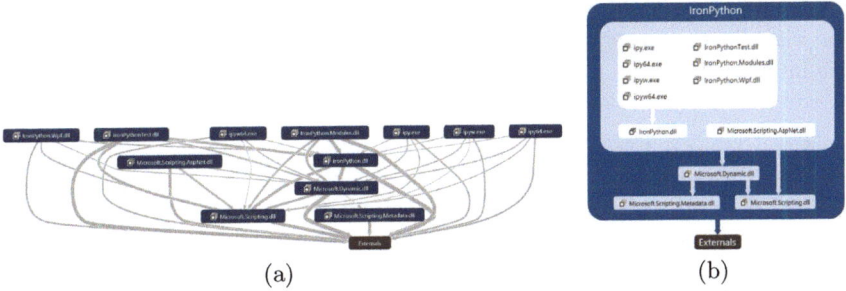

(a)   (b)

Fig. 10.7. Illustration of edge compression to simplify dense graphs. 10.7(a) the top-level component graph produced by the Visual Studio code-dependency analysis tool for the IronPython code base with 39 edges. 10.7(b) a power-graph decomposition of the same graph leaves only six aggregate links without any loss of connectivity information for an 85% compression rate.

Projection Techniques

The Ploceus system [52] is concerned with allowing users a multitude of ways to extract graph views of tabular data. A key part of their system is a *network schema* view of the rich heterogenous graphs that can easily be obtained from such data. The network schema shows a graph of the types of nodes in the network and the types of links that are allowed between pairs of such nodes. The network schema view thus provides a powerful affordance for restricting the set of nodes and edges shown in the actual network visualization. That is, the user can select a subset of the available node and edge types that will appear in the network visualization. An important concept in realizing these final network views is *projection*. That is, for the final subset view to be usefully representative of the original graph, nodes that are to be omitted from the final view must be *spliced* out of the network, rather than simply removed, potentially leaving the graph disconnected even though a transitive relationship exists.

Filtering Techniques

In contrast to aggregation techniques, *graph filtering* selects an appropriately sized subgraph of the input graph, either by stochastic sampling or by deterministic processes. Typically, an importance function measures the relevance of vertices and edges in the graph and only the most relevant objects are kept. Selections can be done by computing additional vertex and edge attributes, e.g., centrality measures, that indicate how important these features are in the graph [48]. Van Ham and Perer [34] use a degree-of-interest function to determine the relevant subgraph for one or more focus points in the graph. This function evaluates both the graph topology and the multivariate graph attributes based on the selected focal vertex. Subsequently, a maximal interest subgraph of specified size is extracted. By interacting with the visualization, users can expand additional parts of the graph that they are interested in.

10.2.2 Visual Mapping

Once a sub-graph is chosen for actual visualization by application of the data reduction techniques above, there are further scalability considerations in the Visual Mapping stage. By "Visual Mapping" we mean the visual representations of data sets as views and/or visual encodings of data items in views. Multiple views can help with orienting the user of the visualization system when the displayed visuals represent only a very small fraction of the total data space. For example, the display may be configured to show an abstract over-view of a large portion of the full graph while a detailed view shows a much more restricted neighborhood but with many more attributes shown on each of the visible nodes and edges [14]. In general, this mapping happens in the encoding and filtering stages of the pipeline. For multiple views, multiple encoding operations coexist, e.g., Ploceus and Attribute Relationship Graphs both have a central graph view that feeds off of all three encoding operations (for nodes, edges and packs) and peripheral views (fed by node or edge encoding operations).

As already discussed in this chapter, various paradigms exist for visual mappings for graphs, the two most widely known being node-link diagrams and matrix views. The limits of scalability for each of these were discussed in Sect. 10.1 while more exotic representations are discussed in Chap. 7. The appeal of node-link diagrams is that it is fairly natural for most people to illustrate related concepts by connecting labels with lines, and—at least while the diagram is simple enough to be unambiguous—for readers of such diagrams to follow transitive paths. By contrast, matrices offer unambiguous representations of very dense graphs (i.e., graphs with a high proportion of edges to nodes). Henry et al. [38] have elegantly demonstrated that hybrid visual mapping may offer the best of both representations. In their NodeTrix system, they use matrices to display dense parts of a large graph, while these matrices are themselves treated as nodes situated within a larger node-link diagram.

Encodings include input on graph element position from the layout feedback loop of the pipeline. Layout operates on the filtered graph subset and changes the position encoding of elements. Complex encoding operations support rich visual mappings, for example: edge centric schemes such as those suggested by Riche et al. [64] control edge curvature based on edge attributes.

In general, there is a trade-off between scalability of layout techniques and the quality of the resultant drawing. For node-link diagrams, algorithms exist that can obtain layouts that may be useful to show the gross structure of an overall graph for thousands (even hundreds of thousands of nodes) in reasonable time [26, 32, 41]. However, for small diagrams—especially when the nodes are not just points but also need to display multiple attributes—there are additional and computationally expensive considerations for layout, for example: avoiding overlaps between node boundaries [15, 28] and minimizing edge-edge and node-edge crossings [59].

10.2.3 View Transformation

The use of data reduction to limit the view to only a small sub-graph—perhaps a sub-graph that is specifically chosen for a particular line of enquiry—necessitates flexible navigation affordances, to allow analysts to easily refocus on different aspects or parts of the graph. We call this type of navigation *view transformation*. For example, Huang et al. [42] developed an early system for exploring an *infinite* graph by browsing just a small neighbourhood at a time. A simple animated spring algorithm enabled incremental layout as nodes are added to or removed from the neighbourhood, and gave the graph smooth transitions. Fisher coined the term "ego-centric views" [23] to describe views of the graph from one particular node's point of view, or from a small neighborhood.

In the pipeline of Fig. 10.5, this is the chief concern of the brushing stage, where encoding can depend on brushing to highlight (un)selected graph elements, while filtering can depend on brushing to elide (un)selected graph elements. The brushing stage also covers view-specific interaction capabilities, such as panning and zooming to navigate a graph coordinate space. In the full pipeline model (and implementation in the Improvise system [82]), operations at *all* stages can depend on navigation and selection visualization parameters controlled throughout a coordinated multiple view visualization.

Overview+detail can be thought of as branching late in the pipeline with different levels of filtering controlling the portions of graph visible in each view, and more encoding to show increased detail (e.g., attributes) of remaining elements. Robert's *Multiform Visualization* [65] idea boils down to multiple pipelines with varying encoding and filtering.

Focus+context techniques also show the most detail around only a small neighborhood, but they endeavor to show this neighborhood in the context of the larger graph. Focus+context graph techniques can be thought of combinations of visual mapping transformations with multiple views that are nested. A compelling and scalable example of this design concept is the *Topological Fisheye* technique of Gansner et al. [29] and another that takes advantage of graphics hardware is proposed by Zinsmaier et al. [96]. In the topological fisheye system a layout is computed for the entire graph, then a combination of spatial and structural clustering techniques are used to show an abridged view of the graph with only gross detail visible. The authors describe navigation in which the user is able to zoom in to show full detail in a small focal region with the abridged contextual structure still visible.

10.3 Studies on Scalability in Graph Visualization

This section provides a summary of the evaluations related to multivariate graph scalability. Some of them are part of work that has been discussed so far. Use Cases are the most popular form of study, but in many works, these cases were designed to demonstrate the proposed technique rather than being a formal evaluation. This section covers some of the case studies, but the focus is on the formal studies, both qualitative (such as interviews) and quantitative (such as controlled experiments).

10.3.1 Data Transformation and Reduction

Wattenberg [81] described the "pilot usage" of "PivotGraph" (please see Sect. 10.2.1 for details) in his paper. These are essentially observations followed by semi-structured interviews after participants have been using the tool for a considerable period. The results are from five analysts who looked for new patterns in their own data using PivotGraph. They are very familiar with the data, which have been analyzed with other tools.

Three multivariate graphs were used in the study: the first one is a transition matrix consisting of 521 states (nodes) and 2,671 transition probabilities (weighted edges). Besides the edge weighting attribute, each node (state) had four associated categorical attributes. The second dataset is the social network among a community of 146 people within a large company. Each person (node) in the community was classified on five dimensions. The last dataset is similar to the second one: it is the communication patterns among employees of a company, with each employee classified according to five different dimensions. The graphs used in the study is not small. For instance the transition matrix graph had 521 nodes and 2,671 edges. However, due to the aggregation technique deployed in PivotGraph, the number of nodes shown in the examples were less than 100 nodes. In that sense, visual complexity is well under control.

The paper provided detailed description of how PivotGraph was used to analyze these datasets, especially what the new findings were and how they were discovered. This provides support for the claim that PivotGraph can help identify new patterns in multivariate graph data. All the participants are very positive about their experience of using the PivotGraph, and they especially liked the feature that allows quick visual comparison between different pair of dimensions (attributes). All the participants wanted to continue using the tool, together with what they were using already. This shows that PivotGraph can be an useful addition to the multivariate graph analysis tool collection.

The Orion system paper [37] includes three use cases: online medical forum discussions, academic collaborations, and software developments. The first use case involves 3 million discussion posts from `MedHelp.org`. The analyst was able to construct network based on edge weight to answer relevant questions. The resulting visualization led to discovery of errors in the dataset and interesting co-occurrence of forum participants on different medical topics. The second use case used the publication information from the ACM Digital Library to visualize the career development of academics. The last use case is based on the Github data and the visualization showed the difference between the followers to the cities where open-source development are most active.

10.3.2 Visual Mapping

Wu and Takatsuka conducted an user study [91] to evaluate the effectiveness of their multivariate networks visualization method that uses Self-Organizing Map (SOM) to improve its layout. Their method tries to find optimal node distance based on not only graph distance but also graph attribute similarity. The evaluation consisted two parts: two use cases and a controlled experiment. The first use case is a student friendship network, with node attribute being the result of two courses. There are 43 nodes and 55 directed edges in total. The results showed that it was possible to achieve good balance between node attribute clustering (measured by "data distortion") and graph drawing aesthetics (measured by "edge crossings") by adjusting their weighs in the SOM function.

The second use case is based on the Krackhardt's high-tech manager advice network [80]. Again, this is a relative small social network with 21 nodes (the managers) but dense connections (190 edges). Each manager has four attributes: Age, Tenure, Position Level, and Department. The results again showed it is possible to achieve a good balance between the attribute clustering and layout aesthetics. The user study compared their method (Fig. 10.8(b)) with a glphy-base one (Fig. 10.8(a)), in which a star glyph is used to show node attributes. It involved 33 participants performing tasks on 7 synthesis multivariate networks. These networks had between 30-50 nodes, 40-70 edges, and 4 or 10 attributes. The tasks included comparing the set of neighbors of two given nodes in terms of their attribute similarity and comparing relationships within the same set of entities. The results showed that the participants spent more time using the glyph-based visualization, which also had lower accuracy.

A study by Cunningham et al. [9] evaluated their method of visualizing multivariate network using 2.5D surfaces, each of which represents a node attribute. They compared their method, GraphScape [92] (Fig. 10.9(a)), to the approach of using node size to show the attribute value (Fig. 10.9(b)). In the first experiment, the participants were asked to select the 20% nodes with the largest attribute value from the visualization of graphs with up to 100 nodes (Fig. 10.9(b)). The results showed that there was no significant difference in accuracy between the two methods, but it took longer to complete the task with the GraphScape. In the second experiment, the participants were asked to determine the average value of a variable for a cluster of nodes

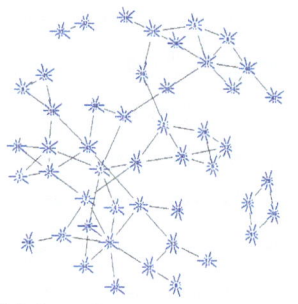
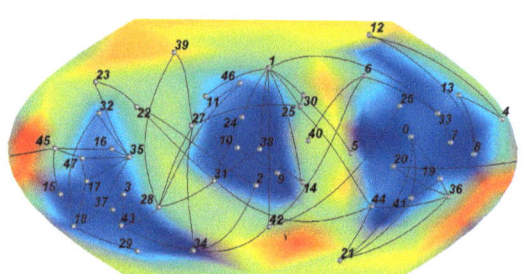

(a) Star Glyph: each node is a star glyph to show its attributes.

(b) Self-Organizing Map-based Hybird Layout: nodes are placed not only to reduce edge crossings but also to show their attribute similarity.

Fig. 10.8. The two multivariate network visualizations used in the user study by Wu and Takatsuka [91] in their paper on hybrid layout method

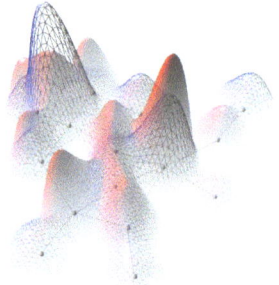
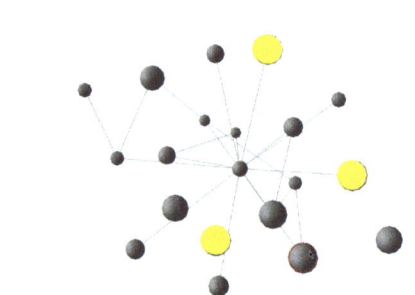

(a) GraphScape: node attributes are shown as 2.5D surfaces. Two attributes are shown as red and blue surface in this example.

(b) The other visualization used in the study: using node size to show attribute value. The task is to select the top 20% nodes with the largest value.

Fig. 10.9. The visualizations used in the evaluation of the GraphScape [92] method

using both visualizations, with similar-sized graphs. The accuracy of GraphScape was found to be significantly greater than that of using node size. However, participants answered significantly faster with node-size visualization, comparing to GraphScape.

There is an increasing usage of curved edges in graph visualization techniques designed to address scalability issues, such as the edge boundling methods discussed in Sect. 10.2.1. There were two user studies [62, 93] on the impact on readability when using curved edges in graph visualization. Edge curvature can be used to encode edge attribute, and it is commonly used in the edge bundling and confluent drawing methods discussed earlier.

The first study consisted of two experiments. The first experiment examine the impact of three different curvature levels on graph readability, with the straight edges (zero curvature) included as the baseline. Participants completed path-finding tasks in a controlled experiment setup, and the graphs used have 20, 50, or 100 nodes. The results showed that using either straight edges or slightly curved edges are more accurate and faster than using heavily curved edges. There was no significant difference in accuracy between the straight edge and the slightly curved edge, but the former is significantly faster. The graph size had a significant impact on speed (each size increase incurred a significant time penalty), but less so on the accuracy. The second experiment included force-directed Lombardi layout [8], which uses circular edges to maximize angular resolution (the minimal angle between edges adjacent to a node). Four tasks were tested in the experiment and the largest graphs have 200 nodes. There was no significant accuracy difference but both straight edge and Lombardi layout were faster than the slightly curved edge. The study by Purchase et al. compared the two variations of the Lombardi layout with straight-edge graphs produced by force-directed method. The size of the graphs used was smaller (20 or 40 nodes) but each size had two edge density level. The three tasks were similar to the second experiment in the study by Xu et al. discussed above. The results were quite different from the previous study: straight edges were found to be faster and more accurate than the two variations of the Lombardi layout. The user preference was also different: Lombardi layout was the choice for aesthetics in this study whereas straight edge was the preferred option in the study by Xu et al.

10.3.3 Navigation and Interaction

Dörk et al. designed a visualization method, PivotPaths [11] (Fig. 10.10), to allow browsing of large data collections through their multiple facets and encourages exploration and serendipitous discoveries. While the method is not designed with multivariate network in mind, it provides a novel way to interactively visualize the relationships between data records through the similarity among their attributes. Because of the design goal, Dörk et al. decided to use a longitudinal study together with observation and semi-structured interviews. The data set used is a collection of academic publications with 160,000 articles, 180,000 authors, and 20,000 keywords. The study was conducted in a research institute with more than 200 recorded user sessions, which were followed by interviews with four participants. The data from the recorded sessions and comments from the interviews confirmed that the PivotPaths provided an integrated view of the three facets in the data (publications, authors, and keywords) and the relationships among them. The participants found the "pivoting" animation is easy to follow and it encouraged them to explore more about the dataset. However, there was someone confusion about pivoting and filtering: some participants expected filtering when they "pivoted" from one facet to another.

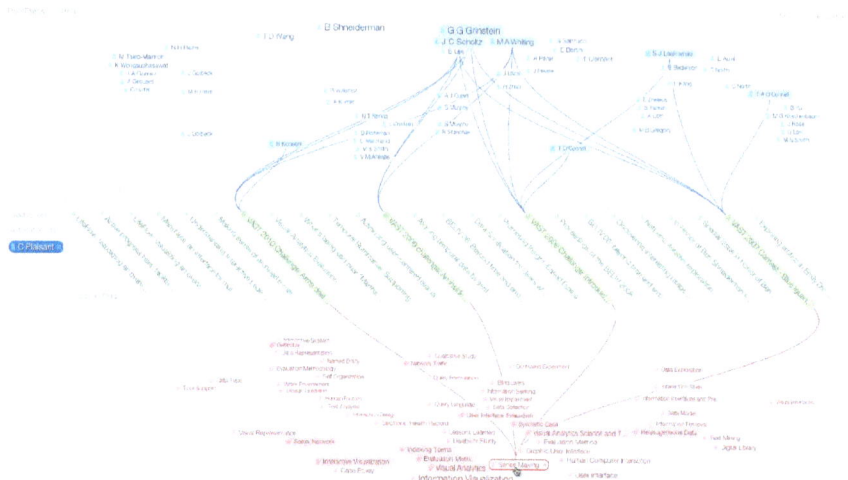

Fig. 10.10. PivotPaths showing the links between the three facets of research publications: author (top), paper (middle), and topic (bottom)

The work by Tomer et al. [56] introduced two navigation techniques, Link Sliding (Fig. 10.11(a)) and Bring & Go (Fig. 10.11(b)), for large networks. Both adopted the focus+context approach and were evaluated in a controlled experiment. The experiment used two randomly generated scale-free graphs, one sparse (1,000 nodes; 1,485 edges), and one dense (1,000 nodes; 2,488 edges). There were 12 participants and the tasks include identifying all nodes connected to a given node, following a link, and returning to a previously visited link (revisit task). Besides the two new techniques, "Pan and Zoom" and "Bird's Eye View" were also included. The Bring & Go technique is significantly faster than Pan-and-Zoom and Bird's-Eye-View in all tasks, but Link Sliding was not significantly faster than Bird's-Eye-View in two out of the three tasks. There was no significant difference in accuracy for the "following" and "revisit" task because there were very few errors. The participants unanimously agreed that Bring & Go was quick, and made the tasks easy. They also found it the least tiring, and most pleasant to use. Link Sliding and Bird's Eye View both received mixed comments regarding the ease and speed at accomplishing tasks. Pan & Zoom was generally rated as slow and difficult to use for the given tasks. Because the two networks used in the experiment shared the same number of nodes, it is not possible to observe how the techniques scale with graph node number. However, edge density did show negative performance impact for some of the tasks. The two techniques were not designed for multivariate networks and may require extra work to provide such support.

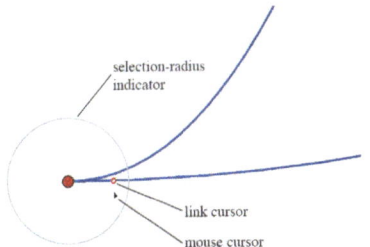 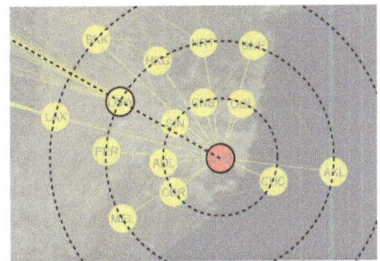

(a) Link Sliding allows the sliding along a (long) edge when the cursor is within the *selection radius*.

(b) Bring & Go makes all the neighbors, some of which normally would be outside the frame, visible within the display.

Fig. 10.11. The Link Sliding and Bring & Go method designed for navigating large graphs

10.4 Challenges and Future Directions

We have attempted to review the state-of-the-art from research and industry in addressing the problem of scalability for multivariate graph visualization; limitations from the hardware and cognitive side were also overviewed. Hopefully, the principles and design guidelines discussed will be useful to implementers of new systems. It should be noted that the systems reviewed here tend to be either research prototypes or visualization platforms specifically designed for a particular type of graph or application. The "holy-grail" of a visualization system that can be easily applied to any type or amount of multivariate graph data remains very much an open challenge. However, the so-called "big-data" problem is ever growing with the steady march of Moores' law and the growth of the internet. Similarly, there is a growing popularity of a network view of data, evidenced by the rise of technologies such as social networking, so-called "graph search", and a move away from tabular data paradigms for storage, such as graph databases. For these reasons we think that more and more researchers and practitioners will begin to explore the use of visualization for very large multivariate graph data and we expect to see rapid developments in this area in the future.

References

1. Abello, J., van Ham, F., Krishnan, N.: ASK-GraphView: A large scale graph visualization system. IEEE Transactions on Visualization and Computer Graphics 12(5), 669–676 (2006)
2. Angles, R., Gutiérrez, C.: Survey of graph database models. ACM Comput. Surv. 40(1) (2008)
3. Auber, D.: Tulip: A huge graph visualisation framework(2003); Mutzel, P., Junger, M. (eds.), http://hal.archives-ouvertes.fr/hal-00307626

4. Berkhin, P.: A survey of clustering data mining techniques. In: Kogan, J., Nicholas, C., Teboulle, M. (eds.) Grouping Multidimensional Data, pp. 25–71. Springer, Heidelberg (2006), http://dx.doi.org/10.1007/3-540-28349-8_2
5. Bezerianos, A., Chevalier, F., Dragicevic, P., Elmqvist, N., Fekete, J.D.: Graphdice: A system for exploring multivariate social networks. Comput. Graph. Forum 29(3), 863–872 (2010)
6. Card, S.K., Mackinlay, J.D., Shneiderman, B.: Readings in information visualization: Using vision to think. Morgan Kaufmann Publishers Inc. (1999)
7. Card, S.K., Mackinlay, J.D., Shneiderman, B. (eds.): Readings in Information Visualization: Using Vision to Think. Morgan Kaufmann (January 1999)
8. Chernobelskiy, R., Cunningham, K.I., Goodrich, M.T., Kobourov, S.G., Trott, L.: Force-directed lombardi-style graph drawing. In: Speckmann, B. (ed.) GD 2011. LNCS, vol. 7034, pp. 320–331. Springer, Heidelberg (2011)
9. Cunningham, A., Xu, K., Thomas, B.H.: Seeing more than the graph – evaluation of multivariate graph visualization methods. In: Proceedings of the Workshop on Interactive Data Exploration and Knowledge Discovery (Part of International Working Conference on Advanced Visual Interfaces 2010), Rome, Italy, pp. 429–429 (2010)
10. Dickerson, M., Eppstein, D., Goodrich, M., Meng, J.: Confluent drawings: Visualizing non-planar diagrams in a planar way. Journal of Graph Algorithms and Applications 9(1), 31–52 (2005)
11. Dörk, M., Riche, N., Ramos, G., Dumais, S.: PivotPaths: strolling through faceted information spaces. IEEE Transactions on Visualization and Computer Graphics 18(12), 2709–2718 (2012)
12. Dwyer, T.: Two-and-a-half-dimensional Visualisation of Relational Networks. Ph.D. thesis, School of Information Technologies, Faculty of Science, University of Sydney (2005)
13. Dwyer, T., Henry Riche, N., Marriott, K., Mears, C.: Edge compression techniques for visualization of dense directed graphs. IEEE Transactions on Visualization and Computer Graphics 19(12), 2596–2605 (2013)
14. Dwyer, T., Marriott, K., Schreiber, F., Stuckey, P., Woodward, M., Wybrow, M.: Exploration of networks using overview+ detail with constraint-based cooperative layout. IEEE Transactions on Visualization and Computer Graphics 14(6), 1293–1300 (2008)
15. Dwyer, T., Marriott, K., Stuckey, P.J.: Fast node overlap removal. In: Healy, P., Nikolov, N.S. (eds.) GD 2005. LNCS, vol. 3843, pp. 153–164. Springer, Heidelberg (2006)
16. Dwyer, T., Mears, C., Morgan, K., Niven, T., Marriott, K., Wallace, M.: Improved optimal and approximate power graph compression for clearer visualisation of dense graphs. In: PacificVis 2014, pp. 105–112. IEEE (2014)
17. Elmqvist, N., Dragicevic, P., Fekete, J.D.: Rolling the dice: Multidimensional visual exploration using scatterplot matrix navigation. IEEE Trans. Vis. Comput. Graph. 14(6), 1148–1539 (2008)
18. Eppstein, D., Goodrich, M.T., Meng, J.Y.: Delta-confluent drawings. In: Healy, P., Nikolov, N.S. (eds.) GD 2005. LNCS, vol. 3843, pp. 165–176. Springer, Heidelberg (2006)
19. Eppstein, D., Goodrich, M.T., Meng, J.Y.: Confluent layered drawings. Algorithmica 47(4), 439–452 (2007)
20. Eppstein, D., Holten, D., Löffler, M., Nöllenburg, M., Speckmann, B., Verbeek, K.: Strict confluent drawing. In: Wismath, S., Wolff, A. (eds.) GD 2013. LNCS, vol. 8242, pp. 352–363. Springer, Heidelberg (2013)

21. Ersoy, O., Hurter, C., Paulovich, F., Cantareiro, G., Telea, A.: Skeleton-based edge bundling for graph visualization. IEEE Transactions on Visualization and Computer Graphics 17(12), 2364–2373 (2011)
22. Fikkert, W., D'Ambros, M., Bierz, T., Jankun-Kelly, T.J.: Interacting with visualizations. In: Kerren, A., Ebert, A., Meyer, J. (eds.) Human-Centered Visualization Environments 2006. LNCS, vol. 4417, pp. 77–162. Springer, Heidelberg (2007)
23. Fisher, D.: Using egocentric networks to understand communication. IEEE Internet Computing 9(5), 20–28 (2005)
24. Foley, J.D., van Dam, A., Feiner, S.K., Hughes, J.F.: Computer Graphics: Principles and Practice in C, 2nd edn. Addison-Wesley (1996)
25. Fortunato, S.: Community detection in graphs. Physics Reports 486(3-5), 75–174 (2010),
http://www.sciencedirect.com/science/article/pii/S0370157309002841
26. Frishman, Y., Tal, A.: Multi-level graph layout on the gpu. IEEE Transactions on Visualization and Computer Graphics 13(6), 1310–1319 (2007)
27. Gansner, E., Hu, Y., North, S., Scheidegger, C.: Multilevel agglomerative edge bundling for visualizing large graphs. In: Proc. PacificVis, pp. 187–194 (2011)
28. Gansner, E.R., Hu, Y.: Efficient node overlap removal using a proximity stress model. In: Tollis, I.G., Patrignani, M. (eds.) GD 2008. LNCS, vol. 5417, pp. 206–217. Springer, Heidelberg (2009)
29. Gansner, E.R., Koren, Y., North, S.C.: Topological fisheye views for visualizing large graphs. IEEE Transactions on Visualization and Computer Graphics 11(4), 457–468 (2005)
30. Ghoniem, M., Fekete, J.D., Castagliola, P.: On the readability of graphs using node-link and matrix-based representations: a controlled experiment and statistical analysis. Information Visualization 4(2), 114–135 (2005)
31. Günnemann, S., Boden, B., Seidl, T.: DB-CSC: A density-based approach for subspace clustering in graphs with feature vectors. In: Gunopulos, D., Hofmann, T., Malerba, D., Vazirgiannis, M. (eds.) ECML PKDD 2011, Part I. LNCS (LNAI), vol. 6911, pp. 565–580. Springer, Heidelberg (2011),
http://dx.doi.org/10.1007/978-3-642-23780-5_46
32. Hachul, S., Jünger, M.: An experimental comparison of fast algorithms for drawing general large graphs. In: Healy, P., Nikolov, N.S. (eds.) GD 2005. LNCS, vol. 3843, pp. 235–250. Springer, Heidelberg (2006)
33. Hadlak, S., Schumann, H., Cap, C.H., Wollenberg, T.: Supporting the visual analysis of dynamic networks by clustering associated temporal attributes. IEEE Transactions on Visualization and Computer Graphics 19(12), 2267–2276 (2013)
34. van Ham, F., Perer, A.: Search, show context, expand on demand: Supporting large graph exploration with degree-of-interest. IEEE Transactions on Visualization and Computer Graphics 15(6), 953–960 (2009)
35. He, B., Fang, W., Luo, Q., Govindaraju, N.K., Wang, T.: Mars: a mapreduce framework on graphics processors. In: Proceedings of the 17th international conference on Parallel architectures and compilation techniques, PACT 2008, pp. 260–269. ACM Press, New York (2008),
http://doi.acm.org/10.1145/1454115.1454152
36. Heer, J., Boyd, D.: Vizster: Visualizing online social networks. In: Proceedings of the IEEE Symposium on Information Visualization (InfoVis), pp. 33–40. IEEE, Minneapolis (October 2005)

37. Heer, J., Perer, A.: Orion: A system for modeling, transformation and visualization of multidimensional heterogeneous networks. In: 2011 IEEE Conference on Visual Analytics Science and Technology (VAST), pp. 51–60. IEEE (2011)
38. Henry, N., Fekete, J.D., McGuffin, M.J.: NodeTrix: A hybrid visualization of social networks. IEEE Transactions on Visualization and Computer Graphics 13(6), 1302–1309 (2007)
39. Holten, D.: Hierarchical edge bundles: Visualization of adjacency relations in hierarchical data. IEEE TVCG 12(5), 741–748 (2006)
40. Holten, D., van Wijk, J.J.: Force-directed edge bundling for graph visualization. Comp. Graph. Forum 28(3), 670–677 (2009)
41. Hu, Y.: Efficient, high-quality force-directed graph drawing. Mathematica Journal 10(1), 37–71 (2005)
42. Huang, M.L., Eades, P., Wang, J.: On-line animated visualization of huge graphs using a modified spring algorithm. Journal of Visual Languages & Computing 9(6), 623–645 (1998)
43. Hui, P., Pelsmajer, M.J., Schaefer, M., Stefankovic, D.: Train tracks and confluent drawings. Algorithmica 47(4), 465–479 (2007)
44. Hurter, C., Ersoy, O., Telea, A.: Graph bundling by kernel density estimation. Comp. Graph. Forum 31(3 pt. 1), 865–874 (2012),
http://dx.doi.org/10.1111/j.1467-8659.2012.03079.x
45. Hurter, C., Ersoy, O., Telea, A.: Smooth bundling of large streaming and sequence graphs. In: Proceedings of the PacificVis 2013 (2013)
46. Hurter, C., Telea, A., Ersoy, O.: Moleview: An attribute and structure-based semantic lens for large element-based plots. IEEE Transactions on Visualization and Computer Graphics 17(12), 2600–2609 (2011),
http://dx.doi.org/10.1109/TVCG.2011.223
47. Hurter, C., Tissoires, B., Conversy, S.: Fromdady: Spreading aircraft trajectories across views to support iterative queries. IEEE Transactions on Visualization and Computer Graphics 15(6), 1017–1024 (2009),
http://dx.doi.org/10.1109/TVCG.2009.145
48. Jia, Y., Hoberock, J., Garland, M., Hart, J.: On the visualization of social and other scale-free networks. IEEE Transactions on Visualization and Computer Graphics 14(6), 1285–1292 (2008)
49. Klippel, A., Hardisty, F., Li, R., Weaver, C.: Colour enhanced star plot glyphs – can salient shape characteristics be overcome? Cartographica 44(3), 217–231 (2009)
50. Klippel, A., Weaver, C., Robinson, A.C.: Analyzing cognitive conceptualizations using interactive visual environments. Cartography and Geographic Information Science 38(1), 52–68 (2011)
51. Lambert, A., Bourqui, R., Auber, D.: Winding roads: Routing edges into bundles. Comp. Graph. Forum 29(3), 432–439 (2010)
52. Liu, Z., Navathe, S.B., Stasko, J.T.: Network-based visual analysis of tabular data. In: Proceedings of the IEEE Conference on Visual Analytics Science and Technology (VAST 2011), pp. 41–50. IEEE (2011)
53. McDonnel, B., Elmqvist, N.: Towards utilizing gpus in information visualization: A model and implementation of image-space operations. IEEE Transactions on Visualization and Computer Graphics 15(6), 1105–1112 (2009),
http://dx.doi.org/10.1109/TVCG.2009.191
54. Miller, G.A.: The magical number seven, plus or minus two: Some limits on our capacity for processing information. The Psychological Review 63(2), 81–97 (1956)

55. Millodot, M.: Dictionary of Optometry and Visual Science. Butterworth-Heinemann (1997)
56. Moscovich, T., Chevalier, F., Henry, N., Pietriga, E., Fekete, J.D.: Topology-aware navigation in large networks. In: Proceedings of the SIGCHI Conference on Human Factors in Computing Systems, pp. 2319–2328. ACM (2009)
57. Nguyen, Q., Eades, P., Hong, S.-H.: On the faithfulness of graph visualizations. In: Didimo, W., Patrignani, M. (eds.) GD 2012. LNCS, vol. 7704, pp. 566–568. Springer, Heidelberg (2013)
58. Owens, J.D., Luebke, D., Govindaraju, N., Harris, M., Krüger, J., Lefohn, A., Purcell, T.J.: A survey of general-purpose computation on graphics hardware. Computer Graphics Forum 26(1), 80–113 (2007), http://www.blackwell-synergy.com/doi/pdf/10.1111/j.1467-8659.2007.01012.x
59. Pupyrev, S., Nachmanson, L., Bereg, S., Holroyd, A.E.: Edge routing with ordered bundles. In: Speckmann, B., van Kreveld, M. (eds.) GD 2011. LNCS, vol. 7034, pp. 136–147. Springer, Heidelberg (2011)
60. Purchase, H.C.: Which aesthetic has the greatest effect on human understanding? In: DiBattista, G. (ed.) GD 1997. LNCS, vol. 1353, pp. 248–261. Springer, Heidelberg (1997)
61. Purchase, H.C., Carrington, D., Allder, J.-A.: Experimenting with aesthetics-based graph layout. In: Anderson, M., Cheng, P.C.H., Haarslev, V. (eds.) Diagrams 2000. LNCS (LNAI), vol. 1889, pp. 498–501. Springer, Heidelberg (2000)
62. Purchase, H.C., Hamer, J., Nöllenburg, M., Kobourov, S.G.: On the usability of lombardi graph drawings. In: Didimo, W., Patrignani, M. (eds.) GD 2012. LNCS, vol. 7704, pp. 451–462. Springer, Heidelberg (2013)
63. Quercini, G., Ancona, M.: Confluent drawing algorithms using rectangular dualization. In: Brandes, U., Cornelsen, S. (eds.) GD 2010. LNCS, vol. 6502, pp. 341–352. Springer, Heidelberg (2011)
64. Riche, N.H., Dwyer, T., Lee, B., Carpendale, S.: Exploring the design space of interactive link curvature in network diagrams. In: Proceedings of the International Working Conference on Advanced Visual Interfaces, pp. 506–513. ACM (2012)
65. Roberts, J.C.: Multiple-View and Multiform Visualization. In: Erbacher, R., Pang, A., Wittenbrink, C., Roberts, J. (eds.) Proceedings of SPIE Visual Data Exploration and Analysis VII, vol. 3960, pp. 176–185 (January 2000)
66. Royer, L., Reimann, M., Andreopoulos, B., Schroeder, M.: Unraveling protein networks with power graph analysis. PLoS computational biology 4(7), e1000108 (2008)
67. Scheepens, R., Willems, N., van de Wetering, H., Andrienko, G., Andrienko, N., van Wijk, J.J.: Composite density maps for multivariate trajectories. IEEE Transactions on Visualization and Computer Graphics 17(12), 2518–2527 (2011), http://dx.doi.org/10.1109/TVCG.2011.181
68. Shadoan, R., Weaver, C.: Visual analysis of higher-order conjunctive relationships in multidimensional data using a hypergraph query system. IEEE Transactions on Visualization and Computer Graphics 19(12), 2070–2079 (2013)
69. Shneiderman, B.: The eyes have it: A task by data type taxonomy for information visualizations. In: Proceedings of the IEEE Symposium on Visual Languages, pp. 336–343. IEEE (1996)
70. Stasko, J., Görg, C., Liu, Z.: Jigsaw: Supporting investigative analysis through interactive visualization. Information Visualization 7(2), 118–132 (2008)
71. Stell, A.J.: Granulation for graphs. In: Freksa, C., Mark, D.M. (eds.) COSIT 1999. LNCS, vol. 1661, pp. 417–432. Springer, Heidelberg (1999)

72. Thompson, C.J., Hahn, S., Oskin, M.: Using modern graphics architectures for general-purpose computing: a framework and analysis. In: Proceedings of the 35th Annual ACM/IEEE International Symposium on Microarchitecture, MICRO, vol. 35, pp. 306–317. IEEE Computer Society Press, Los Alamitos (2002), http://dl.acm.org/citation.cfm?id=774861.774894
73. Tominski, C., Abello, J., Schumann, H.: CGV–an interactive graph visualization system. Computers & Graphics 33(6), 660–678 (2009)
74. Vogel, D., Balakrishnan, R.: Distant freehand pointing and clicking on very large, high resolution displays. In: Proceedings of the 18th Annual ACM Symposium on User Interface Software and Technology (UIST 2005), pp. 33–42. ACM Press, New York (2005)
75. Ward, M.O., Grinstein, G.G., Keim, D.A.: Interactive Data Visualization-Foundations, Techniques, and Applications. A K Peters (2010)
76. Ware, C: Information Visualization: Perception for Design, 2nd edn. Morgan Kaufmann (2004)
77. Ware, C., Bobrow, R.: Supporting visual queries on medium-sized node-link diagrams. Information Visualization 4(1), 49–58 (2005)
78. Ware, C., Mitchell, P.: Visualizing graphs in three dimensions. ACM Trans. Appl. Percept. 5(1), 2:1–2:15 (2008)
79. Ware, C., Purchase, H.C., Colpoys, L., McGill, M.: Cognitive measurements of graph aesthetics. Information Visualization 1(2), 103–110 (2002)
80. Wasserman, S., Faust, K.: Social network analysis: methods and applications. Cambridge University Press, Cambridge (1994)
81. Wattenberg, M.: Visual exploration of multivariate graphs. In: Proceedings of the SIGCHI Conference on Human Factors in Computing Systems, pp. 811–819. ACM (2006)
82. Weaver, C.: Building highly-coordinated visualizations in Improvise. In: Proceedings of the IEEE Symposium on Information Visualization (InfoVis 2004), pp. 159–166. IEEE Computer Society, Austin (October 2004)
83. Weaver, C.: Visualizing coordination in situ. In: Proceedings of the IEEE Symposium on Information Visualization (InfoVis 2005), pp. 165–172. IEEE Computer Society, Minneapolis (October 2005)
84. Weaver, C.: Metavisual exploration and analysis of DEVise coordination in Improvise. In: Proceedings of the International Conference on Coordinated & Multiple Views in Exploratory Visualization (CMV), pp. 79–90. IEEE Computer Society, London (July 2006)
85. Weaver, C.: Cross-filtered views for multidimensional visual analysis. IEEE Transactions on Visualization and Computer Graphics 16(2), 192–204 (2010)
86. Weaver, C.: Multidimensional data dissection using attribute relationship graphs. In: Proceedings of the IEEE Symposium on Visual Analytics Science and Technology (VAST), pp. 75–82. IEEE, Salt Lake City (October 2010)
87. Westheimer, G.: Visual acuity. In: Kaufman, P.L., Alm, A. (eds.) Adler's Physiology of the Eye: Clinical Applications, 10th edn., ch. 17, pp. 453–469. Elsevier (1987)
88. Wikipedia: List of display by pixel density: Apple, http://en.wikipedia.org/wiki/List_of_displays_by_pixel_density#Apple (last accessed November, 2013)
89. Wolfe, J.M.: Guided search 2.0: A revised model of visual search. Psychonomic Bulletin & Review 1(2), 202–238 (1994)
90. Wolfe, J.M., Cave, K.R., Franzel, S.L.: Guided search: An alternative to the feature integration model for visual search. Journal of Experimental Psychology 15(3), 419–433 (1989)

91. Wu, Y., Takatsuka, M.: Visualizing multivariate networks: A hybrid approach. In: Proceedings of the IEEE Pacific Visualization Symposium (PacificVis 2008), pp. 223–230 (2008)
92. Xu, K., Cunningham, A., Hong, S.H., Thomas, B.H.: GraphScape: integrated multivariate network visualization. In: Proceedings of the 6th International Asia-Pacific Symposium on Visualization, Sydney, Australia, Febraury 2007, pp. 33–40 (2007)
93. Xu, K., Rooney, C., Passmore, P., Ham, D.H., Nguyen, P.: A user study on curved edges in graph visualization. IEEE Transactions on Visualization and Computer Graphics 18(12), 2449–2456 (2012)
94. Zhou, Y., Cheng, H., Yu, J.X.: Graph clustering based on structural/attribute similarities. Proc. VLDB Endow. 2(1), 718–729 (2009), http://dl.acm.org/citation.cfm?id=1687627.1687709
95. Zhou, Y., Cheng, H., Yu, J.: Clustering large attributed graphs: An efficient incremental approach. In: 2010 IEEE 10th International Conference on Data Mining (ICDM), pp. 689–698 (2010)
96. Zinsmaier, M., Brandes, U., Deussen, O., Strobelt, H.: Interactive level-of-detail rendering of large graphs. IEEE Transactions on Visualization and Computer Graphics 18(12), 2486–2495 (2012)

Author Index

Abello, James 151
Archambault, Daniel 151

Börner, Katy 175

Diehl, Stephan 13
Dwyer, Tim 207

Elmqvist, Niklas 97

Fekete, Jean-Daniel 97

Gou, Liang 37

Hagen, Hans 175
Holten, Danny 207
Hurter, Christophe 207

Jankun-Kelly, T.J. 207

Kennedy, Jessie 151
Kerren, Andreas 1, 175
Kobourov, Stephen 151
Kohlbacher, Oliver 61, 127

Ma, Kwan-Liu 37, 151
Miksch, Silvia 151
Muelder, Chris 37, 151

Nöllenburg, Martin 207

Pretorius, A. Johannes 77
Purchase, Helen C. 1, 77

Roberts, Jonathan C. 127

Schreiber, Falk 61, 175
Stasko, John T. 77

Telea, Alexandru C. 13, 151

van Wijk, Jarke J. 97
von Landesberger, Tatiana 97

Ward, Matthew O. 1, 61, 127
Weaver, Chris 207
Wybrow, Michael 97

Xu, Kai 207

Yang, Jing 127

Zeckzer, Dirk 175
Zhou, Michelle X. 37, 127
Zimmer, Björn 97

MIX
Papier aus verantwortungsvollen Quellen
Paper from responsible sources
FSC® C105338

If you have any concerns about our products,
you can contact us on
ProductSafety@springernature.com

In case Publisher is established outside the EU,
the EU authorized representative is:
**Springer Nature Customer Service Center GmbH
Europaplatz 3, 69115 Heidelberg, Germany**

Printed by Libri Plureos GmbH
in Hamburg, Germany